Progress in Mathematics
Vol. 17

Edited by
J. Coates and
S. Helgason

Birkhäuser
Boston · Basel · Stuttgart

Samuel M. Vovsi

Triangular Products of Group Representations and Their Applications

1981

Birkhäuser
Boston • Basel • Stuttgart

Author:

Samuel M. Vovsi
Riga Polytechnic Institute
Riga 226355
USSR

LC 81-71610

ISBN 3-7643-3062-7
Printed in USA

INTRODUCTION

The construction considered in these notes is based on a very simple idea. Let (A, G_1) and (B, G_2) be two group representations, for definiteness faithful and finite-dimensional, over an arbitrary field. We shall say that a faithful representation (V, G) is an extension of (A, G_1) by (B, G_2) if there is a G-submodule W of V such that the naturally arising representations (W, G) and $(V/W, G)$ are isomorphic, modulo their kernels, to (A, G_1) and (B, G_2) respectively.

Question. Among all the extensions of (A, G_1) by (B, G_2), does there exist such a "universal" extension which contains an isomorphic copy of any other one?

The answer is in the affirmative. Really, let $\dim A = m$ and $\dim B = n$, then the groups G_1 and G_2 may be considered as matrix groups of degrees m and n respectively. If (V, G) is an extension of (A, G_1) by (B, G_2) then, under certain choice of a basis in V, all elements of G are represented by $(m + n) \times (m + n)$ matrices of the form

$$\left(\begin{array}{c|c} g_1 & 0 \\ \hline h & g_2 \end{array}\right) \qquad (*)$$

where $g_i \in G_i$. It is now clear that the natural representation $(A \oplus B, G^*)$ where G^* is the group of all possible matrices of the form $(*)$ is the desired "universal" extension. Forestalling events, let us say that it is the representation $(A \oplus B, G^*)$ that is called the triangular product of the representations (A, G_1) and (B, G_2). It is the simplest example of this construction which, in general, relates to arbitrary representations over arbitrary commutative rings.

Despite its naturality and transparency, the construction of triangular product appeared in an explicit form only in 1971. It was introduced by Plotkin [50] with the purpose of investigating the semigroup of varieties of group representations. Very soon the construction turned out to have a number of interesting applications in other fields of algebra, and it was successfully used by many authors. Unfortunately, the corresponding results are rather disorderly scattered over the literature, some of them have been published in the editions which are not widely distributed (this also relates to the initial Plotkin's paper), but some of them have not yet been published at all. This explains why a new man wishing to acquire the subject encounters a number of difficulties, not only of mathematical nature.

The aim of the present notes is threefold. First, we try to present a detailed and self-contained account of what have been already done in the area. Second, we hope to persuade the reader that the technique of triangular products may be successfully applied to the investigation of quite concrete and long-familiar algebraic objects, such as augmentation ideals and dimension subgroups, triangular matrix groups and algebras, representations of lattices. Finally, we would like to point out some open questions which seem to be rather interesting.

Although the greater part of the material in these notes is not new, there is a number of places where the existing work has been simplified or generalized. Besides, there are several novel results - for instance, Theorems 9,8, 12.1 and their corollaries, Proposition 8.5.

The paper consists of two chapters. Chapter 1 deals with triangular products themselves, and its results show that this construction is of certain own interest. First, the triangular product is a functor from the category of group representations to itself which, under certain hypotheses, is left or right exact. Further, the triangular product can be defined as a universal object in a special category. It is also of interest that any two triangular decompositions of a given faithful representation have mutually conjugated refinements - the result which to some extent reminds of the classical Krull-Remak-Schmidt Theorem. Finally, the triangular product agrees very well with the multiplication of varieties of group representations: for instance, if ϱ_1 and ϱ_2 are arbitrary representations over a field, then

$$\operatorname{var}(\varrho_1 \triangledown \varrho_2) = \operatorname{var}\varrho_1 \cdot \operatorname{var}\varrho_2.$$

Chapter 2 is devoted to various applications of triangular products. Without going into details, let us note here some typical consequences which follow from the main results of this chapter.

1. The semigroup of T-ideals of an absolutely free associative algebra over a field is free.

2. If K is an integral domain, then the following identity forms a basis for the identities of the full triangular representation $(K^n, T_n(K))$:

$$(1-[x_1, y_1]) \ldots (1-[x_n, y_n]) \qquad \text{if } |K^*| = \infty,$$
$$(1-[x_1, y_1]z_1^m) \ldots (1-[x_n, y_n]z_n^m) \quad \text{if } |K^*| = m < \infty.$$

3. If K is an integral domain, a complete description of the identities of the full triangular group $T_n(K)$ is given; for example, if K is a field of characteristic

0, then these identities are generated by

$$[[x_1, y_1], [x_2, y_2], \dots, [x_n, y_n]].$$

4. Let K be a Dedekind domain, P its field of fractions and F an absolutely free group. If I and J are completely invariant ideals of the group algebra PF, then

$$IJ \cap KF = (I \cap KF)(J \cap KF).$$

5. For every integer $n \geqslant 0$ there exists a finite group G such that the augmentation terminal of the group ring $\mathbb{Z}G$ is not less than $\omega + n$ where ω is the first infinite ordinal.

6. For every group G and every field K

$$\delta_{\omega+1}(G, K) = \delta_{\omega+2}(G, K) = \dots = \delta_{\omega 2}(G, K);$$

here $\delta_\alpha(G, K)$ is the α-th dimension subgroup of G over K. The full description of these subgroups in group theoretic terms is given.

7. If G is a periodic group, then

$$\delta_{\omega+1}(G, \mathbb{Z}) \subseteq \gamma_\omega(G)$$

where $\gamma_\omega(G)$ is the ω-th term of the lower central series of G.

8. If L is a finite distributive lattice, then over an arbitrary field there exists a representation (V, G) such that the lattice of all G-submodules of V is isomorphic to L.

The notes are completed with a brief appendix on triangular products of certain other objects closely related to group representations. It begins with triangular products of representations of associative and Lie algebras. The corresponding definitions, theorems and proofs are, as a rule, completely analogous to their "group" prototypes. In particular, this allows to obtain in a uniform manner certain results on groups, associative algebras and Lie algebras which were earlier proved separately in each concrete situation (note, for instance, the description of the identities of triangular matrix groups and algebras). In conclusion we consider triangular products of linear automata and discuss a little their applications to decomposition problems, which turned out to be rather encouraging.

One object of these notes is to present all of the theory in a form understandable for many people. So, only the standard algebra background (say, Lang's "Algebra") plus a little bit from the theory of varieties of algebraic structures is presupposed of the reader. Furthermore, we also try to avoid long chains of logical cosequences: except Sections 1 and 2, on which the whole paper is based, all the other sections can be read more or less independently according to the following scheme:

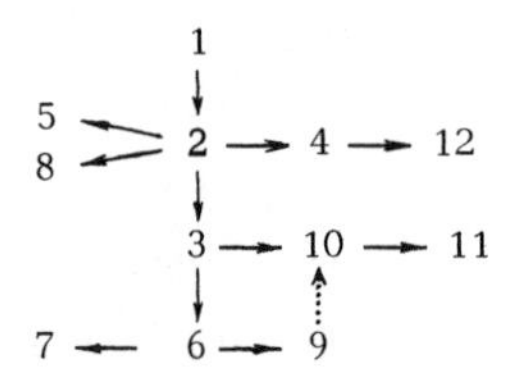

ACKNOWLEDGEMENTS

I am greatly indebted to my teacher B.I.Plotkin who is the founder of the theor considered here, and from whom I gained much of my knowledge. I would like to expr my sincere thanks to G.M.Bergman for his interest in these notes and the encouragement he has given me.

My thanks go also to Mrs.Galina Kravtsova for her efforts and patience in typi this manuscript.

Riga, March 1981 S.M.Vovsi

CONTENTS

Chapter 1

TRIANGULAR PRODUCTS

1.Preliminaries

This introductory section aims to recall briefly some initial notions and facts concerning the category of group representations, closure operations on classes of representations and varieties of group representations which we shall need later. Since the corresponding proofs are standard and straightforward, they are often omitted. For a systematic and detailed discussion of this subject the reader is referred to Plotkin's papers [51] and [55].

Representations. Let K be an arbitrary but fixed commutative ring with identity which will be usually referred to as the basic ring. Suppose there is given a representation of a group G in a (left unitary) K-module A, that is, a homomorphism $\varrho: G \longrightarrow$ $\longrightarrow \mathrm{Aut}_K A$. Denoting for any $a \in A$, $g \in G$

$$ag^{\varrho} = a \circ g,$$

we obtain an action of G on A, that is, a map $(a, g) \longmapsto a \circ g$ from $A \times G$ to A, satisfying the following conditions:

(i) for every $g \in G$ the map $a \longmapsto a \circ g$ is an automorphism of the module A;

(ii) $(a \circ g_1) \circ g_2 = a \circ (g_1 g_2)$ for every $a \in A$, g_1, $g_2 \in G$.

Conversely, let there is given an action $\circ$ of a group G on a K-module A. For every $g \in G$ denote by g^{ϱ} the map $a \longmapsto a \circ g$. It is evident that we obtain a representation $\varrho : G \longrightarrow \mathrm{Aut}\, A$. Thus a representation is actually a triple $(A, G, \circ)$ as well as a triple (A, G, ϱ).

Throughout these notes a representation $\varrho : G \longrightarrow \mathrm{Aut}\, A$ will be denoted by

$\rho = (A, G)$, or simply by (A, G); A is called the domain of action, but G the acting group of the representation ρ. If $\rho = (A, G)$ is any representation, then its kernel $\operatorname{Ker} \rho = \operatorname{Ker}(A, G)$ is the kernel of the corresponding homomorphism $\rho : G \longrightarrow \operatorname{Aut} A$. A representation ρ is faithful if $\operatorname{Ker} \rho = \{1\}$. In this case the group G can be considered as a subgroup of $\operatorname{Aut} A$. A representation $\rho = (A, G)$ is called trivial, or a unit representation if $\operatorname{Ker} \rho = G$. If $A = \{0\}$, ρ is a zero representation.

Let $\rho = (A, G)$ be a representation and $H = \operatorname{Ker} \rho$. Then $G/H = \bar{G}$ acts on A by the rule $a \circ gH = a \circ g$. We obtain a representation $(A, \bar{G}) = \bar{\rho}$ which is already faithful; it is called the faithful image of the representation ρ.

The group algebra of a group G over K is denoted by KG. If (A, G) is a representation, then the action of G on A induces the action of KG on A by the rule: for any $a \in A$, $u = \sum \lambda_i g_i \in KG$

$$a \circ u = \sum \lambda_i (a \circ g_i).$$

Therefore A can be regarded as a right KG-module (when K is clear from the context, it is often said "G-module" instead of "KG-module"). Conversely, a structture of KG-module on A determines an action of G on A, i.e. a representation (A, G).

A group G acts on KG by right multiplication, giving the representation (KG, G), called the regular representation of G over K and denoted by $\operatorname{Reg}_K G = \operatorname{Reg} G$.

The category of representations. Let (A, G) and (B, H) be arbitrary representations over K. A (homo)morphism $\mu : (A, G) \longrightarrow (B, H)$ is a pair, consisting of a homomorphism of K-modules $\mu : A \longrightarrow B$ and a homomorphism of groups $\mu : G \longrightarrow H$ (it is convenient to denote both these maps by a single letter), which satisfies the condition

$$\forall\ a \in A,\ g \in G:\ (a \circ g)^{\mu} = a^{\mu} \circ g^{\mu}.$$

Evidently the class of all representations over K together with all homomorphisms forms a category, called the category of group representations over K and denoted by REP-K. It is easy to understand that the category of all groups is a subcategory of REP-K and that for every group G the category of right KG-modules is a subcategory of REP-K as well.

A homomorphism $\mu: (A, G) \longrightarrow (B, H)$ is a monomorphism (epimorphism, isomorphism) in REP-K if $\mu: A \longrightarrow B$ and $\mu: G \longrightarrow H$ are both monomorphisms (epimorphisms, isomorphisms). If (A, G) is a representation, H a subgroup of G and B an H-submodule of A, then there naturally arises a representation (B, H) which is called a subrepresentation of (A, G). Clearly (B, H) is a subrepresentation of (A, G) if and only if there exists a monomorphism $(B, H) \longrightarrow (A, G)$.

The fact that representations ρ and σ are isomorphic is denoted by $\rho \cong \sigma$. Representations ρ and σ will be called equivalent if their faithful images are isomorphic; this is denoted by $\rho \sim \sigma$. (Note that in the category of representations of a fixed group the equivalency of representations is defined in another way).

Let $\mu: (A, G) \longrightarrow (B, H)$ be a homomorphism and let $A_o = \operatorname{Ker}(A \longrightarrow B)$, $G_o = \operatorname{Ker}(G \longrightarrow H)$. It is easy to see that (A_o, G_o) is a subrepresentation of (A, G). Moreover, this subrepresentation is normal in (A, G), that is, satisfies the following conditions:

(i) $G_o \triangleleft G$;

(ii) A_o is a G_o-submodule of A;

(iii) the induced action of G_o on A/A_o is trivial.

The subrepresentation (A_o, G_o) is called the kernel of μ and is denoted by $\operatorname{Ker} \mu$.

Conversely, let $\rho_o = (A_o, G_o)$ be a normal subrepresentation of $\rho = (A,G)$

(notation: $\rho_o \lhd \rho$). Then the group G/G_o acts on the module A/A_o by the rule

$$(a + A_o) \circ (gG_o) = a \circ g + A_o,$$

and we obtain the factor-representation $\rho / \rho_o = (A/A_o, G/G_o)$. There exists the canonical epimorphism $\varkappa : \rho \longrightarrow \rho/\rho_o$ whose kernel is ρ_o, and usual arguments show that every epimorphic image of ρ can be realized in such a way.

A homomorphism $\mu : (A, G) \longrightarrow (B, H)$ is called right if $\mu : A \longrightarrow B$ is an isomorphism, and left if $\mu : G \longrightarrow H$ is an isomorphism. Without loss of generality we may assume that a right (left) homomorphism acts identically on the left (right) side of a representation. For example, the canonical epimorphism of a representation (A, G) onto its faithful image $(A, G/\mathrm{Ker}\,(A, G))$ is a right epimorphism.

Clearly every right epimorphic image of (A, G) is isomorphic to some factor-representation $(A, G/H)$, where $H \subseteq \mathrm{Ker}\,(A, G)$. Hence, the faithful image of a representation is its "least" right epimorphic image.

Let $\rho_i = (A_i, G_i)$, $i \in I$, be arbitrary representations. Denote by $A = \overline{\prod_{i \in I}} A_i$ the Cartesian sum of A_i, $i \in I$, and by $G = \overline{\prod_{i \in I}} G_i$ the Cartesian product of G_i, $i \in I$. The group G acts on A componentwise and so there arises a representation $\rho = (A, G)$ which is called the Cartesian product of the representations ρ_i and is denoted by

$$\rho = \overline{\prod_{i \in I}} \rho_i = \overline{\prod_{i \in I}} (A_i, G_i).$$

It is easy to see that ρ is the product of the objects ρ_i in REP-K.

Now let us take the direct sum $\bigoplus_{i \in I} A_i$ of modules and the direct product $\prod_{i \in I} G_i$ of groups; the naturally arising representation

$$(\oplus A_i, \prod G_i) = \prod \rho_i = \prod (A_i, G_i)$$

is called the direct product of the representations ρ_i.

Finally, we can also consider the representation $(\oplus A_i, \overline{\prod} G_i)$ with the componentwise action. It is called the semicartesian (right Cartesian) product of the representations ϱ_i.

Closure operations. The next concept which will be important in what follows is the concept of closure operation on classes of representations. It should be mentioned here that the first systematic use of closure operations in algebra appeared in papers by P.Hall (e.g. [19]), although similar ideas occured in earlier papers by Baer and Plotkin.

By a class of group representations $\mathfrak{X}$ we always mean a class with two properties:

(i) $\varrho \in \mathfrak{X}$ and $\varrho \cong \varrho'$ imply $\varrho' \in \mathfrak{X}$,

(ii) $\mathfrak{X}$ contains all zero representations.

The two extreme classes of representations are the class of all representations $\mathfrak{D}$ and the class of zero representations $\mathfrak{E}$. If a representation ϱ is contained in a class $\mathfrak{X}$, we say that ϱ is an $\mathfrak{X}$ -representation.

An operation on classes of group representations is a function U assigning to every class $\mathfrak{X}$ of representations a class $U\mathfrak{X}$ such that

$$\mathfrak{X} \subseteq U\mathfrak{X} \subseteq U\mathfrak{Y}$$

whenever $\mathfrak{X} \subseteq \mathfrak{Y}$. Products of operations are defined by the rule

$$(UV)\mathfrak{X} = U(V\mathfrak{X}).$$

A class $\mathfrak{X}$ is said to be U-closed if $\mathfrak{X} = U\mathfrak{X}$. An operation U is called a closure operation if $U = U^2$; in this case the class $U\mathfrak{X}$ is U-closed for every $\mathfrak{X}$.

Let $\{U_i | i \in I\}$ be a set of operations. Denote by $\langle U_i | i \in I \rangle$ the operation assigning to each $\mathfrak{X}$ the minimal class containing $\mathfrak{X}$ and closed with respect to all U_i. Obviously $\langle U_i | i \in I \rangle$ is a closure operation. Furthermore, if U and V are operations, then $U \leqslant V$ means that $U\mathfrak{X} \subseteq V\mathfrak{X}$ for every $\mathfrak{X}$. The following simple but useful lemma is due to P.Hall [19].

1.1. LEMMA. If U and V are closure operations and $VU \leqslant UV$, then

$\langle U, V\rangle = UV$. ■

Let us define now some concrete closure operations on classes of group representations.

S: $\rho \in S\mathfrak{X}$ if ρ is a subrepresentation of an $\mathfrak{X}$-representation.

Q: $\rho \in Q\mathfrak{X}$ if ρ is an epimorphic image of an $\mathfrak{X}$-representation.

C: $\rho \in C\mathfrak{X}$ if ρ is the Cartesian product of a set of $\mathfrak{X}$-representations.

D and D_o: $\rho \in D\mathfrak{X}$ $(D_o\mathfrak{X})$ if ρ is the direct product of a (finite) set of $\mathfrak{X}$-representations.

C_r: $\rho \in C_r\mathfrak{X}$ if ρ is the semicartesian product of a set of $\mathfrak{X}$-representations.

Q_r: $\rho \in Q_r\mathfrak{X}$ if ρ is a right epimorphic image of an $\mathfrak{X}$-representation.

V: $\rho \in V\mathfrak{X}$ if some right epimorphic image of ρ lies in $\mathfrak{X}$.

R: $(A, G) \in R\mathfrak{X}$ if A is a sum of its KG-submodules A_i such that $(A_i, G) \in \mathfrak{X}$.

S_r: $\rho \in S_r\mathfrak{X}$ if there exists a right monomorphism from ρ to an $\mathfrak{X}$-representation.

1.2. LEMMA. The following relations are valid:

$$QV \leq VQ,\ SV \leq VS,\ CV \leq VC,\ C_rV \leq VC_r,$$
$$SQ \leq QS,\ CQ \leq QC,\ C_rQ \leq QC_r,\ CS \leq SC,$$
$$C_rS \leq SC_r,\ Q_rV \leq VQ_r,\ SQ_r \leq Q_rS,\ CQ_r \leq Q_rC.$$

P r o o f. For example, let us verify the first relation. Suppose $(A, G) \in QV\mathfrak{X}$, then $(A, G) = (B/B_o, H/H_o)$, where $(B, H) \in V\mathfrak{X}$. It follows from the definition of the operation V that there exists a normal subgroup H_1 of H such that $H_1 \subseteq \mathrm{Ker}\,(B, H)$ and $(B, H/H_1) \in \mathfrak{X}$. Then $H_oH_1 \lhd H$ and H_oH_1 acts trivially on B/B_o. Hence there arises a representation $(B/B_o, H/H_oH_1)$. Since $(B, H/H_1) \in \mathfrak{X}$, we have $(B/B_o, H/H_oH_1) \in Q\mathfrak{X}$. The latter representation is a right epimorphic image of $(B/B_o, H/H_o)$, whence $(A, G) = (B/B_o, H/H_o) \in VQ\mathfrak{X}$, as required.

The other relations are proved in a similar manner. ■

From Lemmas 1.1 and 1.2 we obtain:

1.3. COROLLARY. $\langle V, Q, S, C\rangle = VQSC$, $\langle V, Q, S, C_r\rangle = VQSC_r$, $\langle V, Q_r, S, C\rangle = VQ_rSC$. ■

A class of group representations is called saturated if it is V- and Q_r-closed. Clearly a class $\mathfrak{X}$ is saturated if and only if

$$\rho \in \mathfrak{X} \text{ and } \rho \sim \rho' \text{ imply } \rho' \in \mathfrak{X}.$$

If any two representations are equivalent, they both arise from the same faithful representation, that is, from the same action. Therefore, as far as we are concerned with abstract properties of actions, the classes of representation in these notes will be, as a rule, saturated.

Identities and varieties. Given a representation $\rho = (A, G)$ over K, the group algebra KG naturally acts on the module A. Suppose F_∞ is the absolutely free group of countable rank with free generators $x_1, x_2, \ldots, x_n, \ldots$; $u = u(x_1, \ldots, x_n) = \sum \lambda_i f_i(x_1, \ldots, x_n)$ is an element of KF_∞ (where $\lambda_i \in K$, $f_i \in F_\infty$). We say that the bi-identity

$$y \circ u(x_1, \ldots, x_n) = 0$$

is satisfied in the representation ρ if

$$a \circ u(g_1, \ldots, g_n) = 0$$

for any $a \in A$ and any $g_1 \ldots, g_n \in G$. The element $u = u(x_1, \ldots, x_n)$ is called an identity of the representation ρ.

In other words, an element $u \in KF_\infty$ is an identity of ρ if for every homomorphism $F_\infty \longrightarrow G$ the corresponding image of u in KG annihilates A.

If $\mathfrak{X}$ is a class of representations, then an element $u \in KF_\infty$ is called an identity of $\mathfrak{X}$ if it is an identity of every representation from $\mathfrak{X}$.

A class of group representations is called a variety if it consists of all representations satisfying some given set of identities. It is easy to verify that the set of all identities of an arbitrary variety is a two-sided ideal of KF_∞ which is invariant under all endomorphisms of the group F_∞. Such ideals are called completely invariant (= fully invariant = fully characteristic). The usual Galois-type argument shows that there is a natural one-to-one correspondence between the varieties of representations over K and the completely invariant ideals of KF_∞.

The ideal of identities of a variety $\mathfrak{X}$ is denoted by $\mathrm{Id}\,\mathfrak{X}$.

EXAMPLES. 1) The class $\mathfrak{S}$ of all trivial representations is a variety, for it can be determined by a single bi-identity

$$y \circ (1 - x) = 0.$$

The ideal of identities of $\mathfrak{S}$ is generated by all $1 - f$, where $f \in F_\infty$; it is called the augmentation ideal of KF_∞ and is denoted by $\Delta = \Delta_{F_\infty}$.

2) A representation (A, G) is called n-stable if there is a series of G-submodules

$$0 = A_0 \subseteq A_1 \subseteq \dots \subseteq A_n = A$$

such that G acts trivially on every factor A_{n+1}/A_n. The class $\mathfrak{S}^n$ of all n-stable representations is a variety, for it is determined by the bi-identity

$$y \circ (1 - x_1)(1 - x_2) \dots (1 - x_n) = 0;$$

it is clear that $\mathrm{Id}\,\mathfrak{S}^n = \Delta^n$.

3) For any variety of groups Θ denote by $\omega\Theta$ the class of all representations (A, G) such that $G/\mathrm{Ker}\,(A, G) \in \Theta$. This class is a variety since if Θ is determined by a set of group identities $\{f_i\}$, then $\omega\Theta$ is determined by the set $\{1 - f_i\}$. In particular, if Θ is the variety of unit groups, then $\omega\Theta = \mathfrak{S}$.

4) Evidently the class $\mathfrak{D}$ of all representations over K and the class $\mathfrak{E}$ of all zero representations are varieties; they are called trivial varieties. The corresponding completely invariant ideals in KF_∞ are $\{0\}$ and KF_∞ respectively.

1.4. PROPOSITION. If K is a field, then every proper (i.e. $\neq KF_\infty$) completely invariant ideal is contained in the augmentation ideal Δ.

P r o o f. Let I be such an ideal and $u(x_1, \dots, x_n) = \sum \lambda_i f_i(x_1, \dots, x_n) \in I$. Since I is completely invariant,

$$u(1, \dots, 1) = \sum \lambda_i f_i(1, \dots, 1) = \sum \lambda_i \in I.$$

Suppose $\sum \lambda_i \neq 0$. Since K is a field, it follows that $1 \in I$, i.e. $I = KF_\infty$ which is impossible. Therefore $\sum \lambda_i = 0$, and so $I \subseteq \Delta$. ■

EQUIVALENTLY: Every nonzero variety of group representations over a field contains $\mathfrak{S}$. ■

An analogue of the Birkhoff Theorem holds:

1.5. THEOREM. A class of representations is a variety if and only if it is closed under the operations V, Q, S, C. ■

Since the intersection of any collection of varieties is a variety, for any class of representations $\mathfrak{X}$ there exists the least variety containing $\mathfrak{X}$, called the variety generated by $\mathfrak{X}$ and denoted by var $\mathfrak{X}$. It follows from 1.3 and 1.5 that

$$\operatorname{var} \mathfrak{X} = VQSC\mathfrak{X}.$$

Let $\mathfrak{X}$ be a class of representations. A representation ρ is called a residually $\mathfrak{X}$-representation if there is a collection of $\rho_i \lhd \rho \; (i \in I)$ with trivial intersection and with $\rho/\rho_i \in \mathfrak{X}$. In this case we also say that ρ is approximated by $\mathfrak{X}$-representations. The class of residually $\mathfrak{X}$-representations is denoted by $A\mathfrak{X}$. By the Remak Theorem, $A\mathfrak{X}$ coincides with the class of subcartesian products of $\mathfrak{X}$-representations, and so we have the following relation between closure operations:

$$A \leq SC.$$

Therefore any variety is A-closed.

Now let $\mathfrak{X}$ be a variety and $\rho = (A, G)$ an arbitrary representation. If B_i, $i \in I$, are G-submodules of A such that $(A/B_i, G) \in \mathfrak{X}$, then $(A/\cap B_i, G)$ is a residually $\mathfrak{X}$-representation and so belongs to $\mathfrak{X}$. Thus there is the least G-submodule B of A with the property $(A/B, G) \in \mathfrak{X}$. This submodule is called the $\mathfrak{X}$-verbal of ρ and is denoted by

$$\mathfrak{X}^*(\rho) = \mathfrak{X}^*(A, G).$$

1.6. PROPOSITION. Let $\mathfrak{X}$ be a variety.

(i) If $\rho = \prod \rho_i$, then $\mathfrak{X}^*(\rho) = \prod \mathfrak{X}^*(\rho_i)$.

(ii) If $\rho_1 \subseteq \rho_2$, then $\mathfrak{X}^*(\rho_1) \subseteq \mathfrak{X}^*(\rho_2)$.

(iii) If μ is a homomorphism of a representation ρ, then $\mathfrak{X}^*(\rho^\mu) = \mathfrak{X}^*(\rho)^\mu$. ■

Given two classes of group representations $\mathfrak{X}$ and $\mathfrak{Y}$, their product $\mathfrak{X}\mathfrak{Y}$ is defined as follows: a representation (A, G) belongs to $\mathfrak{X}\mathfrak{Y}$ if and only if there is a G-submodule B of A such that $(B, G) \in \mathfrak{X}$ and $(A/B, G) \in \mathfrak{Y}$. Evidently this multiplication is associative.

Now let $\mathfrak{X}$ and $\mathfrak{Y}$ be varieties. It is easily verified that $\mathfrak{X}\mathfrak{Y}$ is also a variety and

$$\mathrm{Id}(\mathfrak{X}\mathfrak{Y}) = \mathrm{Id}\,\mathfrak{X} \cdot \mathrm{Id}\,\mathfrak{Y}.$$

Consequently, the system of all varieties of group representations over a given ring K forms a semigroup $\mathcal{M}(K)$ which is anti-isomorphic to the semigroup of completely invariant ideals of KF_∞.

Free representations. Given two non-empty sets X and Y, take the free group $F = F(X)$ on X and the free KF-module $\Phi = \bigoplus_{y \in Y} yKF$, having the set Y as a basis. It is easy to understand that the representation (Φ, F) is a free object in the category REP-K. It is called the (absolutely) free representation on a pair of sets $\{Y, X\}$. In particular, the regular representation $\mathrm{Reg}\, F = (KF, F)$ is free.

Let $\mathfrak{X}$ be a variety and let $E = \Phi / \mathfrak{X}^*(\Phi, F)$. There naturally arises the representation (E, F) called the free representation on $\{Y, X\}$ of a variety $\mathfrak{X}$. Notice that every variety may be considered as a category, and its free representations are just the free objects of this category.

The rank of the representation (E, F) is the pair of cardinalities $(|Y|, |X|)$. Sometimes $|Y|$ is called the left rank, $|X|$ the right rank of (E, F). If $|Y| = 1$, (E, F) is called the cyclic free representation of rank $|X|$ of the variety $\mathfrak{X}$. The most important is the free cyclic representation of countable rank $(KF_\infty / \mathfrak{X}^*(KF_\infty, F_\infty), F_\infty)$.

1.7. PROPOSITION. Let $(KF_\infty, F_\infty) = \mathrm{Reg}\, F_\infty$ be the absolutely free cyclic representation of countable rank. Then for any variety $\mathfrak{X}$

$$\mathfrak{X}^*(KF_\infty, F_\infty) = \mathrm{Id}\,\mathfrak{X}.$$

P r o o f. Denote $\mathfrak{X}^*(KF_\infty, F_\infty) = A$. If $I = \mathrm{Id}\,\mathfrak{X}$, then it is evident that $(KF_\infty / I, F_\infty) \in \mathfrak{X}$, whence $I \supseteq A$. On the other hand, let $u = u(x_1, \dots, x_n) \in I$. Since $(KF_\infty / A, F_\infty) \in \mathfrak{X}$, u is an identity of this representation. Therefore for elements $\bar{x}_i = x_i + A \in KF_\infty / A$ we have

$$0 = u(\bar{x}_1, \dots, \bar{x}_n) = u(x_1, \dots, x_n) + A,$$

and so $u(x_1, \dots, x_n) \in A$. ∎

The free cyclic representation of countable rank of a variety $\mathfrak{X}$ is denoted by $\mathrm{Fr}\,\mathfrak{X}$. By 1.7,

$$\mathrm{Fr}\,\mathfrak{X} = (KF_\infty / \mathrm{Id}\,\mathfrak{X}, F_\infty).$$

Furthermore, it is clear that

$$\mathrm{var}\,(\mathrm{Fr}\,\mathfrak{X}) = \mathfrak{X}.$$

1.8. PROPOSITION. If $\mathfrak{K}$ is an arbitrary class of representations, then all free representations of $\mathrm{var}\,\mathfrak{K}$ belong to the class $\mathrm{VSC}\,\mathfrak{K}$.

P r o o f. Denote $\mathfrak{X} = \mathrm{var}\,\mathfrak{K}$ and $I = \mathrm{Id}\,\mathfrak{X}$. Let us prove, for instance, that the free cyclic representation $(KF_\infty / I, F_\infty)$ of countable rank belongs to $\mathrm{VSC}\,\mathfrak{K}$. Consider the regular representation (KF_∞, F_∞) and let φ_α, $\alpha \in M$, be all of its homomorphisms into representations of the class $\mathfrak{K}$. Suppose for each $\alpha \in M$ that the kernels of φ_α in KF_∞ and F_∞ are A_α and G_α respectively. Then the representation $(KF_\infty/A_\alpha, F_\infty/G_\alpha)$ is isomorphically embedded in some $\mathfrak{K}$-representation. Let $A = \bigcap_{\alpha\in M} A_\alpha$ and $G = \bigcap_{\alpha\in M} G_\alpha$. Using the Remak Theorem and the relation $SC = AS$ between closure operations, we obtain

$$(KF_\infty/A, F_\infty/G) \in SC\mathfrak{K},$$

whence $(KF_\infty/A, F_\infty) \in VSC\mathfrak{K} \subseteq \mathfrak{X}$. It remains to prove that $A = I$.

Since $(KF_\infty/A, F_\infty) \in \mathfrak{X}$, we have $I \subseteq A$. To prove the reverse inclusion, suppose $u = u(x_1, \ldots, x_n) \in A$. The ideal I consists of all identities of the variety $\mathfrak{X}$, and since $\mathfrak{X} = \mathrm{var}\,\mathfrak{K}$, I consists of all identities of the class $\mathfrak{K}$. Thus it suffices to show that in an arbitrary representation $(B, H) \in \mathfrak{K}$ the bi-identity $y \circ u = 0$ is satisfied. Assume the contrary:

$$\exists\, b \in B;\ h_1, \ldots, h_n \in H : b \circ u(h_1, \ldots, h_n) \neq 0. \qquad (1)$$

Consider the homomorphism $\varphi : (KF_\infty, F_\infty) \longrightarrow (B, H)$ defined as follows: $1^\varphi = b$, where 1 is the unit of the group algebra KF_∞; $x_i^\varphi = h_i$, $i = 1, \ldots, n$, and $x_i^\varphi = 1$ for $i > n$. Then φ is one of the homomorphisms φ_α, $\alpha \in M$. Since A is the intersection of the kernels of all such homomorphisms and $u \in A$, we have $u^\varphi = 0$. But

$$u^\varphi = (1 \circ u)^\varphi = b \circ u(h_1, \ldots, h_n) \neq 0$$

in view of (1). This contradiction shows that $A = I$. ■

Let (E, F) be a free representation of a variety $\mathfrak{X}$. By an obvious reason, its faithful image is called a free faithful representation of $\mathfrak{X}$. In particular, consider the free cyclic representation $\mathrm{Fr}\,\mathfrak{X} = (KF_\infty/I, F_\infty)$. The corresponding free faithful representation is denoted by $\overline{\mathrm{Fr}}\,\mathfrak{X}$; clearly $\overline{\mathrm{Fr}}\,\mathfrak{X} = (KF_\infty/I, F_\infty/H)$ where $H = \{f \in F_\infty \mid f - 1 \in I\}$.

EXAMPLES. 1. Let Θ be a variety of groups and F_Θ its free group of countable rank. Consider the variety of representations $\omega\Theta$. It is easy to see that

$$\overline{Fr}\,(\omega\Theta) = \text{Reg}\, F_\Theta = (KF_\Theta\ , F_\Theta\ ).$$

2. If $\mathfrak{S}^n$ is the variety of n-stable representations, then

$$\overline{Fr}\ \mathfrak{S}^n = (KF_\infty / \Delta^n,\ F_\infty / D_n)$$

where $D_n = \{f \in F \mid f - 1 \in \Delta^n\}$ is the so-called n-th dimension subgroup of F_∞ over K.

CONVENTION. From now on, the free group F_∞ of countable rank will be denoted, for brevity, by F.

Thus we have finished our preliminaries and are able now to turn to the main object of these notes.

2. The definition and basic properties of triangular products

The material of this section is due to Plotkin [50].

Let $\rho_i = (A_i, G_i)$, $i = 1, \dots, n$, be any representations over K. As usual, denote by $\bar{\rho}_i = (A_i, \bar{G}_i)$ the faithful image of the representation ρ_i. The direct product of groups $\prod_{i=1}^n \bar{G}_i$ acts componentwise on the module $V = \bigoplus_{i=1}^n A_i$. Since this action is faithful, we shall regard $\prod \bar{G}_i$ as a subgroup of Aut V. Consider in V the series of submodules

$$0 = V_0 \subseteq V_1 \subseteq \dots \subseteq V_i \subseteq V_{i+1} \subseteq \dots \subseteq V_n = V \tag{1}$$

where $V_i = A_1 \oplus \dots \oplus A_i$, $i = 1, \dots, n$. Denote by Φ the stabilizer (= stability group) of the series (1) in Aut V, that is, the set of all $\varphi \in \text{Aut}\, V$ such that each V_i is invariant under φ and φ acts identically on each V_{i+1}/V_i. An elementary verification shows that the subgroups Φ and $\prod \bar{G}_i$ of Aut V have trivial intersection, and $\prod \bar{G}_i$ normalizes Φ, that is $\Phi \lhd \langle \Phi, \prod \bar{G}_i \rangle$. Therefore the group Aut V contains a semidirect product $\Phi \leftthreetimes \prod \bar{G}_i$, and so we have a faithful representation $(V, \Phi \leftthreetimes \prod \bar{G}_i)$.

Further, the natural epimorphism $\prod G_i \longrightarrow \prod \bar{G}_i$ allows to determine the semidirect product $G = \Phi \lambda \prod G_i$ and the epimorphism $G \longrightarrow \Phi \lambda \prod \bar{G}_i$. The latter allows to "lift" the representation $(V, \Phi \lambda \prod \bar{G}_i)$ up to the representation $\varrho = (V, G)$, called the <u>triangular product of the representations</u> $\varrho_1, \ldots, \varrho_n$ and denoted by

$$\varrho = \varrho_1 \triangledown \ldots \triangledown \varrho_n = \mathop{\triangledown}_{i=1}^{n} \varrho_i$$

or, if the detailed notation is necessary, by

$$(V, G) = (A_1, G_1) \triangledown \ldots \triangledown (A_n, G_n) = \mathop{\triangledown}_{i=1}^{n} (A_i, G_i).$$

Evidently the representations ϱ_i are canonically embedded in ϱ and so can be considered as subrepresentations of ϱ.

Thus, we have given the "outer" definition of the triangular product as an abstract operation on the class of all group representations over K. Now let us give the "inner" definition of this construction. Let $\varrho = (V, G)$ be a representation, and let $\varrho_i = (A_i, G_i)$, $i = 1, \ldots, n$, be a sequence of its subrepresentations. We say that ϱ <u>decomposes into the triangular product of its subrepresentations</u> $\varrho_1, \ldots, \varrho_n$ if the following conditions are satisfied.

(i) The subrepresentation $(V, \langle G_1, \ldots, G_n \rangle)$ of (V, G) decomposes into the direct product of ϱ_i, $i = 1, \ldots, n$; in particular, $V = \bigoplus_{i=1}^{n} A_i$, $\langle G_1, \ldots, G_n \rangle = \prod_{i=1}^{n} G_i$.

(ii) G possesses a normal subgroup Φ, acting faithfully on V, whose natural image in $\mathrm{Aut}\, V$ coincides with the stabilizer of the series $0 = V \subseteq V_1 \subseteq \ldots \subseteq V_n = V$, where $V_i = A_1 \oplus \ldots \oplus A_i$.

(iii) G decomposes into a semidirect product $G = \Phi \lambda \prod_{i=1}^{n} G_i$.

To prove that the "outer" definition is equivalent to the "inner" one, notice first that the "outer" triangular product satisfies, of course, the conditions (i) - (iii). Hence it remains to verify that these conditions determine the representation ϱ uniquely. Beforehand we will establish a technical lemma.

2.1. LEMMA. Consider representations (A, G) and (A', G') where $G = G_1 \lambda G_2$, $G' = G'_1 \lambda G'_2$ and G'_1 acts faithfully on A', and let there be given epimorphisms $\mu : A \twoheadrightarrow A'$, $\mu_1 : G_1 \twoheadrightarrow G'_1$ and $\mu_2 : G_2 \twoheadrightarrow G'_2$. If μ and μ_1 determine an epimorphism $(A, G_1) \twoheadrightarrow (A', G'_1)$, but μ and μ_2 determine an epimorphism $(A, G_2) \twoheadrightarrow (A', G'_2)$, then μ, μ_1 and μ_2 determine an epimorphism $(A, G) \twoheadrightarrow (A', G')$.

P r o o f. First we show that for every $g_1 \in G_1$, $g_2 \in G_2$

$$(g_2^{-1} g_1 g_2)^{\mu_1} = (g_2^{\mu_2})^{-1} g_1^{\mu_1} g_2^{\mu_2} . \tag{2}$$

Both the left and the right parts of (2) belong to G'_1. Therefore, having in mind that G'_1 acts faithfully on A', it suffices to show that

$$x \circ (g_2^{-1} g_1 g_2)^{\mu_1} = x \circ (g_2^{\mu_2})^{-1} g_1^{\mu_1} g_2^{\mu_2}$$

for every $x \in A'$. Let $x = a^{\mu}$ where $a \in A$, then:

$$x \circ (g_2^{-1} g_1 g_2)^{\mu_1} = a^{\mu} \circ (g_2^{-1} g_1 g_2)^{\mu_1} = (a \circ (g_2^{-1} g_1 g_2))^{\mu} ;$$

$$x \circ (g_2^{\mu_2})^{-1} g_1^{\mu_1} g_2^{\mu_2} = ((a^{\mu} \circ (g_2^{-1})^{\mu_2}) g_1^{\mu_1}) g_2^{\mu_2} =$$

$$= ((a \circ g_2^{-1})^{\mu} g_1^{\mu_1}) g_2^{\mu_2} = (a \circ (g_2^{-1} g_1))^{\mu} \circ g_2^{\mu_2} =$$

$$= (a \circ (g_2^{-1} g_1 g_2))^{\mu} .$$

It follows from (2), that μ_1 and μ_2 determine a group epimorphism $G \twoheadrightarrow G'$ which we denote now by μ_3. It remains to show that μ_3 and $\mu : A \twoheadrightarrow A'$ determine an epimorphism of representations $(A, G) \twoheadrightarrow (A', G')$. Let $a \in A$, $g = g_1 g_2 \in G$ where $g_i \in G_i$, then

$$(a \circ g)^{\mu} = ((a \circ g_1) \circ g_2)^{\mu} = (a \circ g_1)^{\mu} \circ g_2^{\mu_2} = a^{\mu} \circ (g_1^{\mu_1} g_2^{\mu_2}) = a^{\mu} \circ g^{\mu_3} ,$$

as required. ■

We can now prove the uniqueness of the triangular product which was claimed above.

2.2. PROPOSITION. The conditions (i) - (iii) determine the representation ϱ uniquely.

P r o o f. Let ϱ decomposes into the triangular product of its subrepresentations $\varrho_1, \ldots, \varrho_n$, but ϱ' into the triangular products of its subrepresentations $\varrho'_1, \ldots, \varrho'_n$. It suffices to prove that

$$(\varrho_1 \cong \varrho'_1) \,\&\, \ldots \,\&\, (\varrho_n \cong \varrho'_n) \Rightarrow (\varrho \cong \varrho').$$

Let $\varrho = (V, G)$, $\varrho' = (V', G')$, $\varrho_i = (A_i, G_i)$ and $\varrho'_i = (A'_i, G'_i)$. Then, by (i) - (iii),

$$\varrho = (\oplus A_i, \Phi \lambda \Pi G_i), \quad \varrho' = (\oplus A'_i, \Phi' \lambda \Pi G'_i).$$

The isomorphisms $\varrho_i \longrightarrow \varrho'_i$ induce the (componentwise) isomorphism $\mu: (V, \Pi G_i) \longrightarrow (V', \Pi G'_i)$. Suppose Ψ is the stabilizer of the series from the definition of ϱ and Ψ' is the analogous stabilizer for ϱ'. By (ii), there exist isomorphisms $\nu : (V,\Phi) \longrightarrow (V,\Psi)$ and $\nu' : (V',\Phi') \longrightarrow (V',\Psi')$, identical on V and V' respectively. It is also clear that that the isomorphism of modules $\mu: V \longrightarrow V'$ naturally induces the isomorphism of representations $\bar{\mu}: (V,\Psi) \longrightarrow (V',\Psi')$.

Thus we have the diagram of isomorphisms

$$\begin{array}{ccc} (V,\Phi) & \dashrightarrow & (V',\Phi') \\ \nu \downarrow & & \downarrow \nu' \\ (V,\Psi) & \xrightarrow{\bar{\mu}} & (V',\Psi') \end{array}$$

Therefore we also have the isomorphism $\nu\bar{\mu}(\nu')^{-1}: (V,\Phi) \longrightarrow (V',\Phi')$ which equals μ on V. Applying Lemma 2.1 to $\nu\bar{\mu}(\nu')^{-1}$ and to $\mu : (V, \Pi G_i) \longrightarrow (V', \Pi G'_i)$, we obtain the required isomorphism $\varrho \longrightarrow \varrho'$. ■

Owing to Proposition 2.2, now we are free not to distinguish the "outer" and the "inner" triangular products.

2.3. PROPOSITION. The triangular product is an associative operation on the class of group representations.

P r o o f. We have to prove that

$$(\rho_1 \nabla \rho_2) \nabla \rho_3 \cong \rho_1 \nabla (\rho_2 \nabla \rho_3). \tag{3}$$

By 2.2, it suffices to show that each part of (3) decomposes into the triangular product of the subrepresentations ρ_1, ρ_2 and ρ_3, i.e. satisfies (i) - (iii). For instance, let us show this for the left part.

Let $\rho_i = (A_i, G_i)$, $i = 1, 2, 3$. Denote the left part of (3) by (V, G), then $(V, G) = (B, H) \nabla \rho_3$, where $(B, H) = \rho_1 \nabla \rho_2$. We have

$$(V, G) = (B \oplus A_3, \Phi \lambda (H \times G_3)), \quad (B, H) = (A_1 \oplus A_2, \Psi \lambda (G_1 \times G_2)),$$

where Φ is the stabilizer of the series $0 \subsetneq B \subsetneq V$, Ψ the stabilizer of the series $0 \subsetneq \subsetneq A_1 \subsetneq B$. It follows that $(V, \langle G_1, G_2, G_3 \rangle) = \prod_{i=1}^{3} \rho_i$.

Further, $G = \Phi \lambda (H \times G_3) = \Phi \lambda ((\Psi \lambda (G_1 \times G_2)) \times G_3) = (\Phi \lambda \Psi) \lambda (G_1 \times G_2 \times G_3)$ and, by the definition of Φ and Ψ, $\Phi \lambda \Psi$ acts faithfully on V. Therefore it remains to verify that $\Phi \lambda \Psi$ (considered as a subgroup of Aut V) coincides with the stabilizer of the series

$$0 \subsetneq A_1 \subsetneq B \subsetneq V. \tag{4}$$

Clearly Φ and Ψ stabilize this series. Conversely, let an element $\sigma \in \mathrm{Aut}\, V$ stabilizes (4). Denote by ψ the automorphism of V defined by the rule

$$a\psi = \begin{cases} a\sigma & \text{if } a \in B, \\ a & \text{if } a \in A_3. \end{cases}$$

Evidently $\sigma \in \Psi$. But $\sigma\psi^{-1}$ acts trivially both an B and on V/B, whence $\sigma\psi^{-1} \in \Phi$. Consequently $\sigma \in \Phi \lambda \Psi$. ■

Proposition 2.3 shows that an arbitrary triangular product can be constructed from "binary" ones. In this connection, we will describe now an important realization of the binary triangular product.

Let $\rho_1 = (A, G_1)$ and $\rho_2 = (B, G_2)$ be arbitrary representations. Take their direct product $(A \oplus B, G_1 \times G_2)$ and consider the additive group $\Phi = \mathrm{Hom}_K(B, A)$. Define an action of the group $G_1 \times G_2$ on Φ. If $\varphi \in \Phi$, $g_1 \in G_1$ and $g_2 \in G_2$, then

$\varphi\circ g_1 g_2$ is defined as follows: for any $b \in B$

$$b(\varphi\circ g_1 g_2) = ((b\circ g_2^{-1})\varphi)\circ g_1.$$

Let G denote the semidirect product corresponding to this action: $G = \Phi\lambda(G_1 \times G_2)$; beforehand, we change the group operation in Φ to multiplicative notation.

Now we define an action of the group G on the module $V = A \oplus B$. For any $a \in \in A$, $b \in B$, $\varphi \in \Phi$, $g_i \in G_i$ set

$$(a + b)\circ \varphi g_1 g_2 = a\circ g_1 + (b\varphi)\circ g_1 + b\circ g_2. \tag{5}$$

A simple verification shows that (5) determines a representation $\varrho = (V, G)$. In particular, let us extract how elements of $G_1 \times G_2$ and Φ act on V. According to (5) we have:

$$(a + b)\circ g_1 g_2 = a\circ g_1 + b\circ g_2, \tag{6}$$

$$a\circ\varphi = a, \quad b\circ\varphi = b\varphi + b. \tag{7}$$

(We emphasize that the action $\circ$ of an element $\varphi \in \Phi \subseteq G$ and the natural action of φ regarded as an element of Hom (B, A) are different things).

It follows from (6) that the direct product $\varrho_1 \times \varrho_2$ is naturally contained in ϱ. It follows from (7) that Φ acts faithfully on V and stabilizes the series $0 \subseteq A \subseteq V$. Finally, it is not hard to see that the image of Φ in Aut V coincides with the stabilizer of this series. Thus the conditions (i) - (iii) are satisfied and so $\varrho = \varrho_1 \triangledown \varrho_2$.

This realization of the triangular product turns out to be very convenient, mainly because it allows us to use the well known properties of the functor Hom.

2.4. PROPOSITION. The kernel of a triangular product is the (direct) product of the kernels of the factors.

P r o o f. Let

$$(V, G) = \mathop{\triangledown}_{i=1}^{n} (A_i, G_i) = (A_1 \oplus \dots \oplus A_n, \Phi\lambda(G_1 \times \dots \times G_n)).$$

Denote Ker $(V, G) = G^o$, Ker $(A_i, G_i) = G_i^o$. Clearly $G_1^o \times \dots \times G_n^o \subseteq G^o$. On the other hand, let $g \in G^o$, then g can be uniquely written in the form $g = \varphi g_1 g_2 \cdots g_n$, where

$\varphi \in \Phi$, $g_i \in G_i$. Take in some A_i an arbitrary element a. Since Φ acts trivially on V_{i+1}/V_i (as before, $V_i = \overset{i}{\underset{j=1}{\oplus}} A_j$, $i = 1,2,\ldots,n$), it follows that $a \circ \varphi = a + b$, where $b \in V_{i-1}$. Therefore $a = a \circ g = (a \circ \varphi) \circ (g_1 \ldots g_n) = (a + b) \circ (g_1 \ldots g_n) = a \circ g_i + b \circ (g_1 \ldots$ $\ldots g_n)$, whence $a = a \circ g_i$, $b \circ (g_1 \ldots g_n) = 0$. Since a is an arbitrary element of A_i, we have $g_i \in G_i^o$. Furthermore,

$$b \circ (g_1 \ldots g_n) = 0 \Longrightarrow b = 0,$$

whence $a \circ \varphi = a$ and $\varphi = 1$. Consequently $g = g_1 \ldots g_n \in G_1^o \times \ldots \times G_n^o$. ■

2.5. COROLLARY. A triangular product is a faithful representation if and only if all its factors are faithful. ■

Our next purpose is to establish certain functor properties of ∇.

2.6. PROPOSITION. Let $\mu : \varrho \longrightarrow \varrho'$ be a homomorphism of representations and let σ be any representation. Then there exists a homomorphism

$$\bar{\mu} : \varrho \nabla \sigma \longrightarrow \varrho' \nabla \sigma$$

which agrees with μ on ϱ and such that:

(i) if μ is a monomorphism, then $\bar{\mu}$ is a monomorphism;

(ii) if μ is an epimorphism and B a projective K-module, then $\bar{\mu}$ is an epimorphism.

P r o o f. We use the Hom-realization of the triangular product described above. Let $\varrho = (A, G_1)$, $\varrho' = (A', G_1')$, $\sigma = (B, G_2)$ and let

$$(V, G) = \varrho \nabla \sigma = (A \oplus B,\ \mathrm{Hom}\,(B, A) \leftthreetimes (G_1 \times G_2)),$$

$$(V', G') = \varrho' \nabla \sigma = (A' \oplus B,\ \mathrm{Hom}\,(B, A') \leftthreetimes (G_1' \times G_2)).$$

Define $\bar{\mu} : V \longrightarrow V'$ as follows:

$$(a + b)^{\bar{\mu}} = a^{\mu} + b$$

for every $a \in A$, $b \in B$. Now, the homorphism of modules $\mu : A \longrightarrow A'$ induces the homomorphism

$$_* : \mathrm{Hom}\,(B, A) \longrightarrow \mathrm{Hom}\,(B, A')$$

in the usual manner: $\varphi^{\mu *} = \varphi\mu$. Therefore we can define a map $\bar{\mu}: G \longrightarrow G'$ by

$$(\varphi g_1 g_2)^{\bar{\mu}} = \varphi^{\mu *} g_1^{\mu} g_2$$

for every $\varphi \in \mathrm{Hom}\,(B, A)$, $g_1 \in G_1$, $g_2 \in G_2$. It is easy to see that this $\bar{\mu}$ is a homomorphism of groups giving, together with $\bar{\mu}: V \longrightarrow V'$, a homomorphism of representations $\bar{\mu}: (V, G) \longrightarrow (V', G')$.

Next, let $(A^o, G_1^o) = \mathrm{Ker}\,\mu$, where $A^o = \mathrm{Ker}\,(A \xrightarrow{\mu} A')$, $G_1^o = \mathrm{Ker}\,(G_1 \xrightarrow{\mu} G_1')$. It is evident that

$$\mathrm{Ker}\,\bar{\mu} = (A^o, \mathrm{Hom}\,(B, A^o) \leftthreetimes G_1^o). \tag{8}$$

This shows that if μ is a monomorphism, $\bar{\mu}$ is a monomorphism as well.

Finally, let μ be an epimorphism and B a projective K-module. Then it is well known that $\mu_* : \mathrm{Hom}\,(B, A) \longrightarrow \mathrm{Hom}\,(B, A')$ is also an epimorphism. It follows immediately that $\bar{\mu}: (V, G) \longrightarrow (V', G')$ is an epimorphism of representations. ∎

2.7. COROLLARY. $\nabla\sigma$ is a covariant left exact functor from REP-K to itself. If the basic ring K is a field, this functor is right exact as well. ∎

2.8. PROPOSITION. If as morphisms we consider only right homomorphisms of representations, then ∇ is a covariant left and right exact functor in both variables.

P r o o f. Let $\mu : \rho \longrightarrow \rho'$ be a right homomorphism. Consider the homomorphism

$$\bar{\mu} : \rho\nabla\sigma \longrightarrow \rho'\nabla\sigma$$

defined as in 2.6. It follows from the definition of $\bar{\mu}$ that it is a right homomorphism. Moreover it is easy to see that if μ is a mono(epi)morphism, then $\bar{\mu}$ is also a mono (epi)morphism.

A simple argument of the same type shows that for any right homomorphisms $\nu : \sigma \longrightarrow \sigma'$ there exists a right homomorphism

$$\bar{\nu} : \rho\nabla\sigma \longrightarrow \rho\nabla\sigma'$$

which is injective (surjective) whenever ν has this property. ∎

Let $(V, G) = \mathop{\nabla}\limits_{i=1}^{n}(A_i, G_i)$ be an arbitrary triangular product. Consider the series

$$0 = V_o \subseteq V_1 \subseteq \ldots \subseteq V_n = V \tag{9}$$

where, as usual, $V_i = A_1 \oplus \ldots \oplus A_i$. It turns out that if K is a field, the lattice of G-submodules of V substantially depends on this series. Namely:

2.9. PROPOSITION. Let K be a field and W a G-submodule of V. Then

$$V_{i-1} \subseteq W \subseteq V_i$$

for some i.

P r o o f. Let V_i be the first term of (9) containing W and let $w \in W \setminus V_{i-1}$. We will prove that $V_{i-1} \subseteq W$. Take an arbitrary $v \in V_{i-1}$ and choose a basis $\{e_k\}$ in V such that

(i) every e_k lies in some A_i,

(ii) $w \in \{e_k\}$ and $v \in \{e_k\}$. Denote by φ the automorphism of V which maps w into $w + v$ and acts identically on all the other basis elements. Evidently φ stabilizes (9), whence $\varphi \in \Phi \subseteq G$. Since W is invariant under G,

$$v = w + v - w = w \circ \varphi - w \in W. \quad ■$$

The series (9) will be called the <u>pivot</u> of the corresponding triangular product. Proposition 2.9 explains the origin of this term.

2.10. COROLLARY. Let K be a field, $(V, G) = (A, G_1) \nabla (B, G_2)$ and W a G-submodule of V. Then either $A \subseteq W$ or $W \subseteq A$. ■

We conclude this section with the following useful lemma.

2.11. LEMMA. Let $(V, G) = (A, G_1) \nabla (B, G_2)$, M a G-submodule of V containing A, and $B_o = B \cap M$. If A is an injective K-module, then there exists a right epimorphism

$$(M, G) \longrightarrow (A, G_1) \nabla (B_o, G_2).$$

P r o o f. We have

$(V, G) = (A \oplus B, \Phi\lambda(G_1 \times G_2))$, $(A, G_1)\nabla(B_o, G_2) = (A \oplus B_o, \Phi_o\lambda(G_1 \times G_2))$ where $\Phi = \mathrm{Hom}(B, A)$, $\Phi_o = \mathrm{Hom}(B_o, A)$. Take any $\varphi \in \Phi$ and denote by φ^{μ} its restriction to B_o. Since A is injective, the map $\mu : \Phi \to \Phi_o$ is an epimorphism. Evidently μ commutes with the action of $G_1 \times G_2$ on Φ, and so we have the epimorphism

$$\mu : G = \Phi\lambda(G_1 \times G_2) \to G_o = \Phi_o\lambda(G_1 \times G_2)$$

identical on $G_1 \times G_2$.

Since $A \subseteq M$, $M = A \oplus (B \cap M) = A \oplus B_o$, whence $(A, G_1)\nabla(B_o, G_2) = (M, G_o)$. Now it is clear that the epimorphism $\mu: G \to G_o$ yields the required right epimorphism $(M, G) \to (M, G_o)$. ■

3. The Embedding Theorem. Connections with closure operations

The main result of this section is a theorem of Plotkin [50] which asserts that, under certain hypotheses, every extension of a representation ϱ_1 by a representation ϱ_2 can be embedded into $\varrho_1 \nabla \varrho_2$. Besides a number of concrete applications, this theorem yields a definition of the triangular product as a universal object in a certain category. In particular, this will enable us to transfer the construction of triangular product to several categories closely related to REP-K: representations of semigroups, rings, algebras, etc. (see Appendix).

3.1. THEOREM (Plotkin [50]). Let (V, G) be a faithful representation, and let A be a direct summand of the K-module V, invariant under G. Denote by (A, G_1) and $(V/A, G_2)$ the faithful images of the naturally arising representations (A, G) and $(V/A, G)$. Then (V, G) is isomorphically embedded into $(A, G_1)\nabla(V/A, G_2)$.

P r o o f. Without loss of generality we may regard G as a subgroup of Aut V. For any $g \in G$ denote by g^{μ_1} the automorphism of A induced by g, and by g^{μ_2}

the corresponding automorphism of V/A. Then μ_1 and μ_2 are epimorphisms of G onto G_1 and G_2 respectively. Let $V = A \oplus B$, $\nu_1 : V \longrightarrow V/A$ the canonical epimorphism, and $\nu_2 : V \longrightarrow B$ the natural projection. The restriction of ν_1 to B is an isomorphism $B \longrightarrow V/A$ which will be also denoted by ν_1, but its inverse $V/A \longrightarrow B$ - by ν_1^{-1}. Then for every $x \in V$ we have

$$(x^{\nu_1})^{\nu_1^{-1}} = x^{\nu_2}.$$

Further, from the representation $(V/A, G_2)$ and ν_1 we obtain a representation (B, G_2) setting for every $b \in B$, $g_2 \in G_2$

$$b \circ g_2 = (b^{\nu_1} \circ g_2)^{\nu_1^{-1}}. \tag{1}$$

Then ν_1 becomes an isomorphism of G_2-modules and, besides that, if $g_2 = g^{\mu_2}$ then

$$b \circ g_2 = (b^{\nu_1} \circ g_2)^{\nu_1^{-1}} = ((b \circ g)^{\nu_1})^{\nu_1^{-1}} = (b \circ g)^{\nu_2}.$$

According to the decomposition $V = A \oplus B$, we can embed G_1 into $\mathrm{Aut}\, V$, the action of G_1 on B being defined trivially. Let $* : G_1 \longrightarrow G_1^*$ be the corresponding embedding. By analogy, acting faithfully on B according to (1), the group G_2 is embedded into $\mathrm{Aut}\, V$; the corresponding embedding is also denoted by $* : G_2 \longrightarrow G_2^* \subseteq \subseteq \mathrm{Aut}\, V$. Then

$$b \circ g_2 = b g_2^* = (b \circ g)^{\nu_2}.$$

Let Φ be the stabilizer of the series $0 \subseteq A \subseteq V$ in $\mathrm{Aut}\, V$. Consider the subgroup $G^* = \langle \Phi, G_1^*, G_2^* \rangle$ of $\mathrm{Aut}\, V$. It is clear that the natural faithful representation (V, G^*) decomposes into the triangular product of its subrepresentations (A, G_1^*) and (B, G_2^*). We will prove that $G \subseteq G^*$. Let $g \in G$, $g_1 = g^{\mu_1}$, $g_2 = g^{\mu_2}$, then $g_1^* \in G_1^*$, $g_2^* \in G_2^*$ and therefore it suffices to show that $g(g_2^*)^{-1}(g_1^*)^{-1} \in \Phi$.

Take $b \in B$. Since $b \circ g_2^* = (b \circ g)^{\nu_2}$, we have $b \circ g - b \circ g_2^* \in A$, whence $b \circ g (g_2^*)^{-1} - b \in A$. Therefore $g(g_2^*)^{-1}$ acts identically on V/A; so does $g(g_2^*)^{-1}(g_1^*)^{-1}$. Now take $a \in A$, then $a \circ g = a \circ g_1^*$ and so

$$a \circ g (g_2^*)^{-1}(g_1^*)^{-1} = (a \circ g_1^*) \circ (g_1^*)^{-1} = a.$$

Hence

$$g(g_2^*)(g_1^*)^{-1} \in \Phi .$$

Thus we have established that $(V, G) \subsetneq (V, G^*) = (A, G_1^*) \triangledown (B, G_2^*)$. Since $(A, G_1^*) \cong (A, G_1)$, $(B, G_2^*) \cong (B, G_2)$, the Theorem follows. ∎

A well known group theoretic theorem of Kaloujnine and Krasner [28] asserts that every extension of a group A by a group B is isomorphic to some subgroup of their wreath product A Wr B. If the basic ring K is a field, Theorem 3.1. (which will be referred to as the Embedding Theorem) turns out to be absolute analogous to the Kaloujnine-Krasner Theorem. Namely, let ρ_1 and ρ_2 be representations. We say that a representation (V, G) is an <u>extension of ρ_1 by ρ_2</u> if there is a G-submodule A of V such that $(A, G) \sim \rho_1$ and $(V/A, G) \sim \rho_2$.

3.2. COROLLARY. Let K be a field. Then every extension of ρ_1 by ρ_2 is equivalent to some subrepresentation of $\rho_1 \triangledown \rho_2$.

P r o o f. Let (V, G) be an extension of ρ_1 by ρ_2, $(A, G) \sim \rho_1$ and $(V/A, G) \sim \rho_2$. Since K is a field, A is a direct summand of V, so that the Embedding Theorem may be applied. Denote by $(V, \bar{G})$, (A, G_1) and $(V/A, G_2)$ the faithful images of (V, G), (A, G) and $(V/A, G)$ respectively. By 3.1.

$$(V, \bar{G}) \subsetneq (A, G_1) \triangledown (V/A, G_2) \cong \bar{\rho}_1 \triangledown \bar{\rho}_2 ,$$

but the last representation is the faithful image of $\rho_1 \triangledown \rho_2$. Therefore there is a subrepresentation of $\rho_1 \triangledown \rho_2$ whose faithful image is isomorphic to $(V, \bar{G})$. ∎

Now let us have a look at the Embedding Theorem from the point of view of category theory. Let ρ_1 and ρ_2 be representations over an arbitrary K. According to the above definition, an extension (V, G) of ρ_1 by ρ_2 may be regarded as a group G together with a short exact sequence of KG-modules

$$0 \longrightarrow A \xrightarrow{\varkappa} V \xrightarrow{\sigma} B \longrightarrow 0 \tag{2}$$

such that $(A, G) \sim \rho_1$, $(B, G) \sim \rho_2$. We denote such an extension by

$$E = (A \overset{\varkappa}{\rightarrowtail} V \overset{\sigma}{\twoheadrightarrow} B, G).$$

The extension E will be called an s-extension if the sequence (2) splits over K (but it need not split over KG!). E is called faithful if the corrposponding representation (V, G) is faithful. Clearly from any extension $E = (A \overset{\varkappa}{\rightarrowtail} V \overset{\sigma}{\twoheadrightarrow} B, G)$ one can pass to its faithful image $\bar{E} = (A \overset{\varkappa}{\rightarrowtail} V \overset{\sigma}{\twoheadrightarrow} B, \bar{G})$, where $\bar{G} = G/\mathrm{Ker}\,(V, G)$.

Given two extensions

$$E = (A \overset{\varkappa}{\rightarrowtail} V \overset{\sigma}{\twoheadrightarrow} B, G) \text{ and } E' = (A' \overset{\varkappa'}{\rightarrowtail} V' \overset{\sigma'}{\twoheadrightarrow} B', G'),$$

a morphism $\alpha : E \longrightarrow E'$ is a four-tuple $\alpha = (\alpha_1, \alpha_2, \alpha_3, \alpha_4)$, where $\alpha_1 : A \longrightarrow A'$, $\alpha_2 : V \longrightarrow V'$ and $\alpha_3 : B \longrightarrow B'$ are homomorphisms of K-modules such that the diagram

$$\begin{array}{ccccccccc} 0 & \longrightarrow & A & \xrightarrow{\varkappa} & V & \xrightarrow{\sigma} & B & \longrightarrow & 0 \\ & & \downarrow{\scriptstyle\alpha_1} & & \downarrow{\scriptstyle\alpha_2} & & \downarrow{\scriptstyle\alpha_3} & & \\ 0 & \longrightarrow & A' & \xrightarrow{\varkappa'} & V' & \xrightarrow{\sigma'} & B' & \longrightarrow & 0 \end{array}$$

commutes, but $\alpha_4 : G \longrightarrow G'$ is a homomorphism of groups satisfying the conditions

$$(a \circ g)^{\alpha_1} = a^{\alpha_1} \circ g^{\alpha_4}, \quad (v \circ g)^{\alpha_2} = v^{\alpha_2} \circ g^{\alpha_4}, \quad (b \circ g)^{\alpha_3} = b^{\alpha_3} \circ g^{\alpha_4}.$$

These morphisms allow us to speak of the category of extensions over K, denoted by $\mathcal{E}(K) = \mathcal{E}$.

To avoid unessential details, from now on in this section we shall consider only faithful extensions. This convention actually does not make any limitation since in the present paper we can always "smoothly" pass from an arbitrary representation to its faithful image and vice versa.

Now let ϱ_1 and ϱ_2 be fixed faithful representations. Note that if A and B are the domains of action of ϱ_1 and ϱ_2 respectively, we can consider any extension of ϱ_1 by ϱ_2 to have a form

$$E = (A \overset{\varkappa}{\rightarrowtail} V \overset{\sigma}{\twoheadrightarrow} B, G) \tag{3}$$

(A and B are fixed!). Define now a special subcategory $\mathcal{E}(\varrho_1, \varrho_2)$ of $\mathcal{E}$. Its objects are faithful s-extensions of ϱ_1 by ϱ_2, but its morphisms are defined as above morphisms $(\alpha_1, \alpha_2, \alpha_3, \alpha_4)$ such that $\alpha_1 = 1_A$, $\alpha_3 = 1_B$. In other words, a morphism from the extension (3) to an extension $E' = (A \overset{\varkappa'}{\rightarrowtail} V' \overset{\sigma'}{\twoheadrightarrow} B, G')$ is a homomorphism of the corresponding representations

$$\alpha : (V, G) \longrightarrow (V', G')$$

such that the diagram

$$\begin{array}{ccccccccc} 0 & \longrightarrow & A & \xrightarrow{\varkappa} & V & \xrightarrow{\sigma} & B & \longrightarrow & 0 \\ & & \| & & \downarrow \alpha & & \| & & \\ 0 & \longrightarrow & A & \xrightarrow{\varkappa'} & V' & \xrightarrow{\sigma'} & B & \longrightarrow & 0 \end{array}$$

commutes.

Evidently the triangular product $\rho_1 \triangledown \rho_2$ can be considered as an object of $\mathcal{E}(\rho_1, \rho_2)$ since $\rho_1 \triangledown \rho_2$ is a faithful extension of ρ_1 by ρ_2 which splits over K. Moreover, a glance over the Embedding Theorem and its proof shows that now it can be reformulated in the following form.

3.3. THEOREM. $\rho_1 \triangledown \rho_2$ is a universal object of $\mathcal{E}(\rho_1, \rho_2)$. The morphism from an arbitrary object of $\mathcal{E}(\rho_1, \rho_2)$ to $\rho_1 \triangledown \rho_2$ is a monomorphism. ∎

Theorem 3.3 gives, in fact, another definition of the triangular product – now in terms of the theory of categories.

Our next objective is to show that the triangular product agrees very well with many closure operations on classes of representations.

Let $\mathfrak{X}_1$ and $\mathfrak{X}_2$ be classes of representations. The triangular product $\mathfrak{X}_1 \triangledown \mathfrak{X}_2$ is the class of all representations $\rho_1 \triangledown \rho_2$, where $\rho_1 \in \mathfrak{X}_1$ and $\rho_2 \in \mathfrak{X}_2$.

3.4. LEMMA [69 , 71] . Let $\mathfrak{X}_1$ and $\mathfrak{X}_2$ be classes of representations over an arbitrary K. Then:

(a) $\mathfrak{X}_1 \triangledown \mathfrak{X}_2 \subseteq (V\mathfrak{X}_1)(V\mathfrak{X}_2)$;

(b) $(S\mathfrak{X}_1) \triangledown \mathfrak{X}_2 \subseteq S(\mathfrak{X}_1 \triangledown \mathfrak{X}_2)$;

(c) $(V\mathfrak{X}_1) \triangledown \mathfrak{X}_2 \subseteq V(\mathfrak{X}_1 \triangledown \mathfrak{X}_2)$;

(f) $\mathfrak{X}_1 \triangledown (V\mathfrak{X}_2) \subseteq V(\mathfrak{X}_1 \triangledown \mathfrak{X}_2)$;

(g) $\mathfrak{X}_1 \triangledown (Q_r\mathfrak{X}_2) \subseteq Q_r(\mathfrak{X}_1 \triangledown \mathfrak{X}_2)$;

(h) $\mathfrak{X}_1 \triangledown (S_r\mathfrak{X}_2) \subseteq S_r(\mathfrak{X}_1 \triangledown \mathfrak{X}_2)$;

(d) $(Q_r \mathfrak{X}_1) \nabla \mathfrak{X}_2 \subseteq Q_r(\mathfrak{X}_1 \nabla \mathfrak{X}_2)$; (i) $\mathfrak{X}_1 \nabla (D\mathfrak{X}_2) \subseteq RV(\mathfrak{X}_1 \nabla \mathfrak{X}_2)$;

(e) $(C\mathfrak{X}_1) \nabla \mathfrak{X}_2 \subseteq A(\mathfrak{X}_1 \nabla \mathfrak{X}_2)$; (g) $\mathfrak{X}_1 \nabla (C_r \mathfrak{X}_2) \subseteq RV(\mathfrak{X}_1 \nabla \mathfrak{X}_2)$.

P r o o f. 1) To prove (a), let $(A, G_1) \in \mathfrak{X}_1$, $(B, G_2) \in \mathfrak{X}_2$ and $(V, G) = (A, G_1) \nabla (B, G_2)$. Take the subrepresentation (A, G) of (V, G) and the corresponding factor-representation $(V/A, G)$. Obviously there exist right epimorphisms $(A, G) \longrightarrow (A, G_1)$ and $(V/A, G) \longrightarrow (B, G_2)$. Therefore $(A, G) \in V\mathfrak{X}_1$, $(V/A, G) \in V\mathfrak{X}_2$, and $(V, G) \in (V\mathfrak{X}_1)(V\mathfrak{X}_2)$.

2) From 2.6(i) it follows (b). From 2.8 we obtain (c), (d), (f), (g) and (h).

3) We prove (e). Let ϱ_i, $i \in I$, be representations from $\mathfrak{X}_1$, $\varrho = \overline{\prod_i} \varrho_i$, and let $\sigma \in \mathfrak{X}_2$. Using the same argument as in the proof of Proposition 2.6, it is easy to see that the canonical projection $\pi_i : \varrho \longrightarrow \varrho_i$ extends to the epimorphism

$$\overline{\pi}_i : \varrho \nabla \sigma \longrightarrow \varrho_i \nabla \sigma .$$

Let $(A^o, G^o) = \operatorname{Ker} \pi_i$. By (8) in Section 1,

$$\operatorname{Ker} \overline{\pi}_i = (A^o, \operatorname{Hom}(B, A^o) \curlywedge G^o)$$

where B is the domain of action of σ. Hence, since $\bigcap_i \operatorname{Ker} \pi_i$ is trivial, $\bigcap_i \operatorname{Ker} \overline{\pi}_i$ is also trivial. This shows that $\varrho \nabla \sigma$ is approximated by $\varrho_i \nabla \sigma$, $i \in I$. Since $\varrho_i \nabla \sigma \in \mathfrak{X}_1 \nabla \mathfrak{X}_2$, we have $\varrho \nabla \sigma \in A(\mathfrak{X}_1 \nabla \mathfrak{X}_2)$.

4) To prove (i), take any $(A, G) \in \mathfrak{X}_1$ and $(B, H) \in D\mathfrak{X}_2$. The latter means that $(B, H) = \prod_i (B_i, H_i)$ where $(B_i, H_i) \in \mathfrak{X}_2$. Denote $(V, T) = (A, G) \nabla (B, H)$ and show that $(V, T) \in RV(\mathfrak{X}_1 \nabla \mathfrak{X}_2)$. Consider in $V = A \oplus (\oplus_i B_i)$ the submodules $V_i = A \oplus B_i$, $i \in I$. It is evident that every V_i is invariant under T, and so there is a subrepresentation

$$(V_i, T) = (A \oplus B_i, \operatorname{Hom}(B, A) \curlywedge (G \times H))$$

of (V, T). Denote

$$(V_i, T_i) = (A, G) \nabla (B_i, H_i) = (A \oplus B_i, \operatorname{Hom}(B_i, A) \curlywedge (G \times H_i)).$$

Since B_i is a direct summand of B, the injection $B_i \longrightarrow B$ induces the epimorphism $\operatorname{Hom}(B, A) \longrightarrow \operatorname{Hom}(B_i A)$. This epimorphism together with the natural projection

$H \longrightarrow H_i$ and identical maps $V_i \longrightarrow V_i$, $G \longrightarrow G$ yields a right epimorphism of representations

$$(V_i, T) \longrightarrow (V_i, T_i).$$

Therefore, since $(V_i, T_i) \in \mathfrak{X}_1 \triangledown \mathfrak{X}_2$, it follows that $(V_i, T) \in V(\mathfrak{X}_1 \triangledown \mathfrak{X}_2)$ for each i. The module V is a sum of T-submodules V_i, whence $(V, T) \in RV(\mathfrak{X}_1 \triangledown \mathfrak{X}_2)$.

The proof of (j) coincides with that of (i). ■

If the basic ring K is a field, we can establish some more inclusions of the type (a) - (j).

3.5. LEMMA [69, 71]. Let $\mathfrak{X}_1$ and $\mathfrak{X}_2$ be classes of representations over a field K. Then, apart from (a) - (j) of the previous lemma, we have:

(k) $\mathfrak{X}_1 \mathfrak{X}_2 \subseteq VSQ_r(\mathfrak{X}_1 \triangledown \mathfrak{X}_2)$;

(l) $(Q\mathfrak{X}_1) \triangledown \mathfrak{X}_2 \subseteq Q(\mathfrak{X}_1 \triangledown \mathfrak{X}_2)$;

(m) $\mathfrak{X}_1 \triangledown (Q\mathfrak{X}_2) \subseteq QS(\mathfrak{X}_1 \triangledown \mathfrak{X}_2)$;

(n) $\mathfrak{X}_1 \triangledown (S\mathfrak{X}_2) \subseteq Q_r S(\mathfrak{X}_1 \triangledown \mathfrak{X}_2)$.

P r o o f. (k) Suppose $(V, G) \in \mathfrak{X}_1 \mathfrak{X}_2$ and $(V, \bar{G}')$ is its faithful image. Pick a G-submodule A of V such that $(A, G) \in \mathfrak{X}_1$ and $(V/A, G) \in \mathfrak{X}_2$. Since K is a field, A has a direct complement in V. By the Embedding Theorem, $(V, \bar{G}')$ is embedded isomorphically in

$$\varrho = (A, G_1) \triangledown (V/A, G_2),$$

(A, G_1) and (B, G_2) being faithful images of (A, G) and $(V/A, G)$ respectively.

Consider $\varrho_1 = (A, G) \triangledown (V/A, G)$. Clearly $\varrho_1 \in \mathfrak{X}_1 \triangledown \mathfrak{X}_2$ and, by 2.8, there exists a right epimorphism of ϱ_1 on ϱ. Taking all of this into account, we have:

$$\varrho_1 \in \mathfrak{X}_1 \triangledown \mathfrak{X}_2 \Longrightarrow \varrho \in Q_r(\mathfrak{X}_1 \triangledown \mathfrak{X}_2) \Longrightarrow (V, \bar{G}) \in SQ_r(\mathfrak{X}_1 \triangledown \mathfrak{X}_2) \Longrightarrow$$
$$\Longrightarrow (V, G) \in VSQ_r(\mathfrak{X}_1 \triangledown \mathfrak{X}_2).$$

(l) A straightforward corollary from 2.6(ii).

(m) Let $(A, G) \in \mathfrak{X}_1$, $(B, G_2) \in \mathfrak{X}_2$, and let $\mu : (B, G_2) \longrightarrow (B', G_2')$ be an epimorphism. Denote $(V, G) = (A, G_1) \triangledown (B, G_2)$, $(V', G') = (A, G_1) \triangledown (B', G_2')$. If Φ_o is the set of all elements of $\Phi = \mathrm{Hom}(B, A)$ annihilating the kernel of the epimorphism $\mu: B \longrightarrow B'$, it is easy to verify that $\Phi_o \triangleleft G$ and that there exists an epimorphism $(V, \Phi_o \lambda (G_1 \times G_2)) \longrightarrow (V', G')$ which agrees with μ on (B, G_2). Since $(V, G) \in \mathfrak{X}_1 \nabla \mathfrak{X}_2$, it follows that $(V, \Phi_o \lambda (G_1 \times G_2)) \in S(\mathfrak{X}_1 \nabla \mathfrak{X}_2)$, whence $(V', G') \in$ $\in QS(\mathfrak{X}_1 \nabla \mathfrak{X}_2)$.

(n) Suppose $(A, G_1) \in \mathfrak{X}_1$, $(B, G_2) \in \mathfrak{X}_2$, and (B', G_2') is a subrepresentation of (B', G_2'). We have to prove that $(A, G_1) \nabla (B', G_2') \in Q_r S(\mathfrak{X}_1 \nabla \mathfrak{X}_2)$. Denote

$$(V, G) = (A, G_1) \nabla (B, G_2) = (A \oplus B, \Phi \lambda (G_1 \times G_2)),$$

$$(V', G') = (A, G_1) \nabla (B', G_2') = (A \oplus B', \Phi' \lambda (G_1 \times G_2')),$$

where $\Phi = \mathrm{Hom}\,(B, A)$, $\Phi' = \mathrm{Hom}\,(B', A)$. We shall assume V' to be naturally contained in V. Evidently V' is invariant under $\Phi \lambda (G_1 \times G_2')$, and therefore there is a subrepresentation $(V', \Phi \lambda (G_1 \times G_2'))$ of (V, G). Since $(V, G) \in \mathfrak{X}_1 \nabla \mathfrak{X}_2$, it follows that $(V', \Phi \lambda (G_1 \times G_2')) \in S(\mathfrak{X}_1 \nabla \mathfrak{X}_2)$.

Define now a map

$$\mu: (V', \Phi \lambda (G_1 \times G_2')) \longrightarrow (V', G') \tag{4}$$

as follows: μ acts identically on V', G_1 and G_2', but for any $\varphi \in \Phi$ its image φ^{μ} is the restriction of φ to B'. Since K is a field, $\mu: \Phi \longrightarrow \Phi'$ is an epimorphism.

It is easy to see that the map (4) is a right epimorphism. Therefore $(V', G') \in$ $\in Q_r S(\mathfrak{X}_1 \nabla \mathfrak{X}_2)$, as required. ∎

4. Generalized triangular products

In the preceding sections we considered only triangular products of finite ordered sets of representations. However, the finitness of the number of factors turns out to be unessential. As has been shown in a paper of Vovsi [73], the construction of triangular product can be naturally generalized to arbitrary partially ordered sets of representations (a particular case of totally ordered sets of representations was considered earlier in a paper by Gringlaz and Plotkin [14]).

In this section we will present several results of [73]. Various applications of these results will be discussed in Chapter 2. As usual, K denotes an arbitrary commutative ring with identity.

Let Λ be a partially ordered (p.o.) set and let for every $\alpha \in \Lambda$ there is given a representation $\varrho_\alpha = (A_\alpha, G_\alpha)$. We set $(V, H) = \prod_{\alpha\in\Lambda} \varrho_\alpha$, that is,

$$V = \bigoplus_{\alpha\in\Lambda} A_\alpha, \quad H = \prod_{\alpha\in\Lambda} G_\alpha$$

and H acts on A componentwise. For any $\alpha \in \Lambda$ define in V two submodules

$$V_\alpha = \bigoplus_{\lambda \leq \alpha} A_\lambda \quad \text{and} \quad V_\alpha^\flat = \bigoplus_{\lambda < \alpha} A_\lambda$$

(if α is a minimal element of Λ, then $V_\alpha^\flat = 0$). Then the set consisting of all V_α and $V_\alpha^\flat$ $(\alpha \in \Lambda)$ is partially ordered ordered with respect to inclusion: if $\alpha < \beta$, then

$$V_\alpha^\flat \subsetneq V_\alpha \subseteq V_\beta^\flat \subsetneq V_\beta .$$

Denote by $\widetilde{\Phi}$ the stabilizer in Aut V of the system of factor-modules $\{V_\alpha/V_\alpha^\flat \mid \alpha \in \Lambda\}$ and by Φ the subset of all $\varphi \in \widetilde{\Phi}$ satisfying the following <u>restrictness property</u>: φ acts identically on all but a finite number direct summands A_α. Clearly $\widetilde{\Phi}$ and Φ are subgroups of Aut V.

Let $(V, \bar{H})$ be the faithful image of (V, H). Considering $\bar{H}$ as a subgroup of Aut V, it is easy to see that $\widetilde{\Phi} \cap \bar{H} = 1$, $\widetilde{\Phi} \lhd \langle \widetilde{\Phi}, \bar{H} \rangle$, and $\Phi \lhd \langle \Phi, \bar{H} \rangle$. Therefore the subgroups $\langle \widetilde{\Phi}, \bar{H} \rangle$ and $\langle \Phi, \bar{H} \rangle$ of Aut V split into semidirect products

$$\langle \widetilde{\Phi}, \bar{H} \rangle = \widetilde{\Phi} \leftthreetimes \bar{H}, \quad \langle \Phi, \bar{H} \rangle = \Phi \leftthreetimes \bar{H}.$$

Thus we have faithful representations

$$(V, \widetilde{\Phi} \leftthreetimes \bar{H}) \quad \text{and} \quad (V, \Phi \leftthreetimes \bar{H}). \tag{1}$$

The canonical epimorphism $H \longrightarrow \bar{H}$ leads to semidirect products $\widetilde{G} = \widetilde{\Phi} \leftthreetimes H$ and $G = \Phi \leftthreetimes H$ and induces epimorphisms $\widetilde{\mu}: G \longrightarrow \widetilde{\Phi} \leftthreetimes \bar{H}$ and $\mu: G \longrightarrow \Phi \leftthreetimes \bar{H}$. By means of $\widetilde{\mu}$ and μ, the representations (1) can be "lifted" up to the representations

$$\widetilde{\varrho} = (V, \widetilde{G}) = (V, \widetilde{\Phi} \leftthreetimes H) \quad \text{and} \quad \varrho = (V, G) = (V, \Phi \leftthreetimes H),$$

called the <u>complete</u> and <u>restricted triangular products</u> of ϱ_α, $\alpha \in \Lambda$, respectively, and denoted

$$\widetilde{\varrho} = \widetilde{\nabla}_{\alpha\in\Lambda} \varrho_\alpha \quad \text{and} \quad \varrho = \nabla_{\alpha\in\Lambda} \varrho_\alpha . \tag{2}$$

Evidently, if Λ is a finite p.o. set, the representations $\tilde{\varrho}$ and ϱ coincide. It is also evident that if Λ is a finite linearly ordered set, then we obtain the usual triangular product $\overset{n}{\underset{i=1}{\nabla}} \varrho_i$ in the sense of Section 2.

Suppose all the representations ϱ_α, $\alpha \in \Lambda$, are isomorphic to a single representation σ. Then, instead of (2), we write $\tilde{\varrho} = \tilde{\nabla}\sigma^\Lambda$ and $\varrho = \nabla\sigma^\Lambda$. In this case $\tilde{\varrho}$ and ϱ are called the complete and restricted triangular Λ-power of σ, respectively.

By analogy with Section 2, let us give an "inner" definition of the triangular product. Let $\tilde{\varrho} = (V, \tilde{G})$ be a representation, Λ a p.o. set, and let $\varrho_\alpha = (A_\alpha, G_\alpha)$, $\alpha \in \Lambda$, be subrepresentations of $\tilde{\varrho}$. We say that $\tilde{\varrho}$ decomposes into the complete triangular product of its subrepresentations ϱ_α, $\alpha \in \Lambda$, if the following conditions are satisfied.

(i) Let $H = \langle G_\alpha \mid \alpha \in \Lambda \rangle$. Then the subrepresentation (V, H) of $\tilde{\varrho}$ decomposes into the direct product of ϱ_α, $\alpha \in \Lambda$.

(ii) $\tilde{G}$ possesses a normal subgroup $\tilde{\Phi}$ acting faithfully on V, whose natural image in Aut V coincides with the stabilizer of the system $\{V_\alpha / V_\alpha^\flat \mid \alpha \in \Lambda\}$, where $V_\alpha = \underset{\lambda \leq \alpha}{\oplus} A_\lambda$, $V_\alpha^\flat = \underset{\lambda < \alpha}{\oplus} A_\lambda$ for each $\alpha \in \Lambda$.

(iii) $\tilde{G} = \tilde{\Phi} \leftthreetimes H$.

An "inner" definition of the restricted triangular product and the proof that both definitions are equivalent are left to the reader.

We shall establish several properties of the above constructions. Fix a non-empty p.o. set Λ and a set of representations $\varrho_\alpha = (A_\alpha, G_\alpha)$, $\alpha \in \Lambda$, and let

$$\tilde{\varrho} = \underset{\alpha \in \Lambda}{\tilde{\nabla}} \varrho_\alpha = (V, \tilde{G}), \quad \varrho = \underset{\alpha \in \Lambda}{\nabla} \varrho_\alpha = (V, G).$$

The results of the present section relate to these fixed representations.

Note, first, that the kernels of both $\tilde{\varrho}$ and ϱ coincide with the direct product of the kernels of ϱ_α. This follows immediately from the definitions. In particular, the triangular product of faithful representations is faithful as well. The following statement is also obvious.

4.1. LEMMA. If Ω is a subset of Λ (with the induced ordering), then

$\tilde{\nabla}_{\alpha\in\Omega} \rho_\alpha$ $(\nabla_{\alpha\in\Omega} \rho_\alpha)$ is isomorphically embedded in $\tilde{\rho}$ (in ρ). ■

A segment of a p.o. set Λ is a non-empty subset Ω such that if $\omega_1, \omega_2 \in \Omega$ and $\omega_1 < \lambda < \omega_2$, then $\lambda \in \Omega$. Special kinds of segments are:

a) a lower segment, i.e. such that $\omega \in \Omega$ & $\lambda < \omega \Rightarrow \lambda \in \Omega$;

b) an upper segment, i.e. such that $\omega \in \Omega$ & $\lambda > \omega \Rightarrow \lambda \in \Omega$.

For any non-empty subset Ω of Λ put

$$V_\Omega = \bigoplus_{\alpha\in\Omega} A_\alpha , \quad H_\Omega = \prod_{\alpha\in\Omega} H_\alpha .$$

4.2. LEMMA. If Ω is a lower segment of Λ, then

$$(V/V_\Omega , \tilde{G}) \sim \tilde{\nabla}_{\alpha\in\Lambda\setminus\Omega} \rho_\alpha , \quad (V/V_\Omega , G) \sim \nabla_{\alpha\in\Lambda\setminus\Omega} \rho_\alpha .$$

P r o o f. (Since Ω is a lower segment, it is easy to see that V_Ω is invariant under G. Therefore the above representations have meaning). We will consider only the restricted product; the complete case is proved in a similar way. Put for brevity $\Omega' = \Lambda\setminus\Omega$. We have

$$(V, G) = (\bigoplus_{\alpha\in\Lambda} A_\alpha , \Phi\lambda H), \quad \nabla_{\alpha\in\Omega'} \rho_\alpha = (\bigoplus_{\alpha\in\Omega'} A_\alpha , \Phi_o \lambda H_o).$$

Having denoted the latter by (V_o, G_o), define an epimorphism $\mu: (V, G) \rightarrow (V_o, G_o)$ as follows: $\mu: V \rightarrow V_o$ and $\mu: H \rightarrow H_o$ are the natural projections, but $\mu: \Phi \rightarrow \Phi_o$ is the restriction of elements of $\Phi \subseteq \mathrm{Aut}\, V$ to the submodule V_o.

A trivial verification shows that these three maps yield an epimorphism $\mu: (V, G) \rightarrow (V_o, G_o)$. The left kernel of μ is V_Ω, so that μ can "pass" through $(V/V_\Omega , G)$. Therefore there exists a right epimorphism

$$\mu_1 : (V/V_\Omega , G) \rightarrow (V_o, G_o),$$

whence $(V/V_\Omega , G) \sim (V_o, G_o)$. ■

From now on in this section, the ring K is a field and all modules are vector spaces over K. Denote by π_α the natural projection of $V = \bigoplus A_\alpha$ on the direct summand A_α.

4.3. LEMMA. Let W be a G-subspace of V. If $W^{\pi_\alpha} \neq 0$, then $V_\alpha^\flat \subseteq W$.

P r o o f. Since $V_\alpha^b = \bigoplus_{\lambda<\alpha} A_\lambda$, it suffices to show that $\lambda<\alpha \Rightarrow A_\lambda \subseteq W$. Since $W^{\pi_\alpha} \neq 0$, there exists $w \in W$ such that

$$w = a_1 + a_2 + \dots + a_n.$$

where $0 \neq a_i \in A_{\alpha_i}$, $\alpha_i \neq \alpha_j$ for $i \neq j$, and $\alpha = \alpha_i$ for some i. Let $\lambda < \alpha$ and $b \in \in A_\lambda$. Choose a basis $\{e_\beta\}$ in V containing all a_i and such that every e_β belongs to some A_γ. Let φ be the automorphism of V for which

$$a_\alpha^\varphi = a_\alpha + b \text{ and } e_\beta^\varphi = e_\beta \quad \text{if } e_\beta \neq a_\alpha.$$

Clearly $\varphi \in \Phi$ and $w \circ \varphi = w + b$. Since W is invariant under G, we have $w \circ \varphi \in W$ and so $b = w \circ \varphi - w \in W$. Therefore $A_\lambda \subseteq W$, as required. ■

4.4. COROLLARY. If all ρ_α, $\alpha \in \Lambda$, are nonzero representations, then for any G-subspace W of V the set

$$\Omega = \{\alpha \mid W^{\pi_\alpha} \neq 0\}$$

is a lower segment of Λ. ■

The next result gives a characterization of invariant subspaces in $\tilde{\rho}$ and ρ. For any subset Ω of Λ denote by Ω_1 the set of all maximal elements of Ω, and let $\Omega_0 = \Omega \setminus \Omega_1$. Then, in particular, $V_\Omega = V_{\Omega_0} \oplus V_{\Omega_1}$, $H_\Omega = H_{\Omega_0} \times H_{\Omega_1}$.

4.5. PROPOSITION. For a subspace W of V the following conditions are equivalent:

(i) W is invariant under $\tilde{G}$;

(ii) W is invariant under G;

(iii) there exists a lower segment Ω of Λ such that $W = V_{\Omega_0} \oplus B$, where B is an H_{Ω_1}-invariant subspace of V_{Ω_1}.

P r o o f. Evidently (i) $\Longrightarrow$ (ii) and (iii) $\Longrightarrow$ (i). To prove (ii) $\Longrightarrow$ (iii), consider in Λ a subset

$$\Omega = \{\lambda \mid \exists \alpha \in \Lambda\, [W^{\pi_\alpha} \neq 0 \ \& \ \lambda \leq \alpha]\}$$

Clearly Ω is a lower segment. Two possibilities may occur.

1) Ω does not possess maximal elements, i.e. $\Omega = \Omega_o$. Then it follows easily from 4.3 that $\lambda \in \Omega \Longrightarrow A_\lambda \subseteq W$. Therefore

$$V_\Omega = \bigoplus_{\lambda \in \Omega} A_\lambda \subseteq W,$$

while the reverse inclusion is obvious. Thus $W = V_\Omega = V_{\Omega_o}$.

2) Ω possesses maximal elements. Then for each $\alpha \in \Omega_o$ there exists $\beta \in \Omega$ for which $\alpha < \beta$. By 4.3, $W \supseteq V_{\Omega_o}$. Since $W \subseteq V_\Omega$, it follows that

$$W = V_{\Omega_o} + B, \quad \text{where } B \subseteq V_{\Omega_1}.$$

Here B must be invariant under H_{Ω_1}, for otherwise W cannot be invariant under G. ■

It should be mentioned that this result contains Proposition 2.9 as a special case: indeed, the set of indices $\{1,2,\ldots, i\}$ in the latter plays the role of the lower segment Ω.

Further, recall that ∇ is an associative operation on the class of group representations (Proposition 2.3). We note in passing that this fact can also be generalized to arbitrary (finite or infinite) restricted triangular products. Namely, by a <u>segmentation</u> of Λ we mean a presentation of Λ as a union of non-empty segments Λ_i, $i \in I$, such that:

1) if $i \neq j$, then $\Lambda_i \cap \Lambda_j = \emptyset$;

2) if $i \neq j$, $\alpha \in \Lambda_i$, $\beta \in \Lambda_j$, and $\alpha < \beta$, then each element of Λ_i is less than each element of Λ_j.

Clearly such a segmentation induces a partial ordering on I, so that I can also be a considered to be partially ordered.

4.6. THEOREM (associativity of generalized triangular products). Let $\Lambda = \bigcup_{i \in I} \Lambda_i$ be a segmentation of Λ and $\varrho^i = \nabla_{\alpha \in \Lambda_i} \varrho_\alpha$. Then

$$\varrho \cong \nabla_{i \in I} \varrho^i.$$

For the proof see [73]. We remark that the analogous assertion for complete triangular products does not hold.

In [43] McLain has introduced a method of constructing locally nilpotent groups which have many interesting and useful properties. We conclude this section by showing that McLain's groups may be naturally defined in terms of triangular products.

Recall the definition of McLain's groups. Let Λ be a linearly ordered set and K a field, and let V be a vector space over K having as a basis a set of elements

$$\{v_\lambda \mid \lambda \in \Lambda\}.$$

For each pair of elements λ and μ from Λ such that $\lambda > \mu$, we define a linear transformation $e_{\lambda\mu}$ of V by the rule

$$v_\lambda \circ e_{\lambda\mu} = v_\mu \quad \text{and} \quad v_\nu \circ e_{\lambda\mu} = 0 \quad \text{if} \quad \nu \neq \lambda.$$

If $a \in K$, the linear transformation $1 + ae_{\lambda\mu}$ of V is nonsingular and

$$(1 + ae_{\lambda\mu})^{-1} = 1 - ae_{\lambda\mu}.$$

<u>The McLain group</u> $M(\Lambda, K)$ is the subgroup of $GL(V)$ generated by all the $1 + ae_{\lambda\mu}$:

$$M(\Lambda, K) = \langle 1 + ae_{\lambda\mu} \mid \lambda, \mu \in \Lambda, \lambda > \mu, \ a \in K \rangle.$$

It is obvious from the definition that each element of $M(\Lambda, K)$ has a unique representation as a finite sum

$$1 + \sum_{\lambda > \mu} a_{\lambda\mu} e_{\lambda\mu} \tag{3}$$

where $a_{\lambda\mu} \in K$, and, conversely, that every element of the form (3) belongs to $M(\Lambda, K)$.

Now let $(K, 1)$ be the one-dimensional representation of the trivial group and let

$$(V, G) = \nabla(K, 1)^{\Lambda}.$$

From the above, it is absolutely clear that $G = M(\Lambda, K)$. This gives another definition of McLain's groups.

NOTE. It is worth mentioning that the first example of McLain's groups (with $\Lambda = \mathbb{Q}$ and $K = \mathbb{Z}_p$) has been given by Ado [1].

5. Isomorphisms and automorphisms of triangular products

Corollary 3.2 shows that certain properties of triangular products are analogous to those of wreath products of groups. Let us emphasize now an essential difference. In this connection, recall a well known theorem of P.Neumann [46] asserting that

$$A \operatorname{wr} B \cong A' \operatorname{wr} B' \Longrightarrow A \cong A' \text{ and } B \cong B'$$

for arbitrary groups A, B, A' and B'. Does the same hold for triangular products? More exactly, does the implication

$$\varrho_1 \triangledown \varrho_2 \cong \varrho_1' \triangledown \varrho_2' \Longrightarrow \varrho_1 \cong \varrho_1' \text{ and } \varrho_2 \cong \varrho_2' \qquad (1)$$

hold for arbitrary representations ϱ_1, ϱ_2, ϱ_1' and ϱ_2' ?

This question is worth studying only if the basic ring K is a field and all the representations in (1) are faithful, for otherwise one can easily show trivial counterexamples. Therefore from now on in this section, K is a field and all representations are faithful.

However, even under these assumptions the question is solved in the negative. Indeed, because of associativity of the triangular product we have

$$(\varrho_1 \triangledown \varrho_2) \triangledown \varrho_3 \cong \varrho_1 \triangledown (\varrho_2 \triangledown \varrho_3),$$

but, in general, $\varrho_1 \triangledown \varrho_2 \ncong \varrho_1$. Now, there naturally arises another question: does the implication (1) hold if all the representations are indecomposable into a nontrivial triangular product? Our purpose is to answer this question in the affirmative.

In fact we will obtain a more general result. First let us give several simple definitions. Let $\varrho = (V, G)$ be a representation and let $g \in G$. A pair of maps

$$v \longmapsto v \circ g, \quad x \longmapsto g^{-1} x g \quad (\text{where } v \in V, \ x \in G)$$

determines an automorphism of ϱ, called the inner automorphism induced by g. If $\sigma = (W, H)$ is a subrepresentation of ϱ, then its image $(W \circ g, g^{-1}Hg)$ under this automorphism is denoted by $\sigma^g = (W, H)^g = (W^g, H^g)$.

Suppose there are given two triangular decompositions of a representation ϱ :

$$\varrho = \varrho_1 \nabla \varrho_2 \nabla \cdots \nabla \varrho_m, \qquad (2)$$

$$\varrho = \varrho_1' \nabla \varrho_2' \nabla \cdots \nabla \varrho_n'. \qquad (3)$$

These decompositions are said to be isomorphic (conjugated) if $m = n$ and there exists an automorphism (inner automorphism) of ϱ which maps ϱ_i onto ϱ_i', $i = 1, 2, \dots, n$.

The decomposition (3) is called a refinement of the decomposition (2) if there is a sequence of integers $0 = n_o < n_1 < n_2 < \dots < n_m = n$ such that

$$\varrho_i = \varrho'_{n_{i-1}+1} \nabla \varrho'_{n_{i-1}+2} \nabla \cdots \nabla \varrho'_{n_i}, \quad i = 1,2,\dots, m.$$

Now we are able to state the main result of this section.

5.1. THEOREM (Vovsi [72]). Any two triangular decompositions of a faithful representation over a field have conjugated refinements.

In particular, this answers the above question.

5.2. COROLLARY. If $\varrho_1 \nabla \dots \nabla \varrho_m \cong \varrho_1' \nabla \dots \nabla \varrho_n'$ where all the ϱ_i and ϱ_j' are ∇-indecomposable, then $m = n$ and $\varrho_i \cong \varrho_i'$ for all $i = 1, \dots, n$.

Proof of Theorem 5.1. Let $\varrho = (V, G)$ be an arbitrary faithful representation over a field K for which there are given two triangular decompositions. Without loss of generality, we shall consider G to be naturally contained in $\mathrm{Aut}\, V$. The proof divides into two stages.

Stage 1. Both the decompositions of (V, G) consist of two factors:

$$(V, G) = (A, G_1) \nabla (B, G_2) = (A \oplus B, \Phi\lambda(G_1 \times G_2)), \qquad (4)$$

$$(V, G) = (A', G_1') \nabla (B', G_2') = (A' \oplus B', \Phi' \lambda(G_1' \times G_2')). \qquad (5)$$

Here, as usual, Φ and Φ' are the stabilizers of the series $0 \subseteq A \subseteq V$ and $0 \subseteq A' \subseteq V$, respectively, in $\mathrm{Aut}\, V$. Since A and A' are invariant under G, it follows from 2.10 that either $A \subseteq A'$ or $A' \subseteq A$. Suppose, for instance, the first, and let Ψ be the stabilizer of the series

$$0 \subseteq A \subseteq A' \subseteq V \qquad (6)$$

in Aut V.

<u>(i) $\Phi \subseteq \Psi$, $\Phi' \subseteq \Psi$, $\Psi \triangleleft G$.</u> Note, first, that the truth of $\Phi \subseteq \Psi$ and $\Phi' \subseteq \Psi$ is evident. Pick a basis $\{e_i\}$ in V which agrees with (6), and let $\psi \in \Psi$. Define φ_1, $\varphi_2 \in \operatorname{Aut} V$ as follows:

$$e_i \circ \varphi_1 = \begin{cases} e_i & \text{if } e_i \in A, \\ e_i \circ \psi & \text{if } e_i \in A' \setminus A, \\ e_i & \text{if } e_i \in A'; \end{cases} \qquad e_i \circ \varphi_2 = \begin{cases} e_i & \text{if } e_i \in A', \\ e_i \circ \psi & \text{if } e_i \notin A'. \end{cases}$$

It is easy to see that $\varphi_1 \in \Phi$, $\varphi_2 \in \Phi'$, $\psi = \varphi_1 \varphi_2$. Therefore $\psi \in G$, whence $\Psi \subseteq G$. Since all the members of (6) are invariant under G, but the stabilizer is always normal in the "normalizer", we have $\Psi \triangleleft G$, as required.

Denote $A' \cap B = C$. By the modular law,

$$A' = A' \cap V = A' \cap (A \oplus B) = A \oplus (A' \cap B) = A \oplus C,$$

whence

$$V = A' \oplus B' = A \oplus C \oplus B'. \tag{7}$$

Let π be the projections of $V = A \oplus B$ on B:

$$\forall\ a \in A,\ b \in B:\ (a + b)^{\pi} = b. \tag{8}$$

Since $\operatorname{Ker} \pi = A$ and $A \cap B' = 0$, the restriction of π to B' is one-to-one. Hence $B'^{\pi} = D$ is a subspace of B isomorphic to B'.

<u>(ii) $B = C \oplus D$.</u> By the definition, C and D are contained in B. Suppose $b \in B$ and $b = a' + b'$, where $a' \in A'$, $b' \in B'$. It follows from the above that $a' = a + c$, $b' = a_1 + d$, where $a, a_1 \in A$, $c \in C$, $d \in D$. Therefore $b = (a + a_1) + (c + d)$, and since $a + a_1 \in A$, $c + d \in B$, we have $b = c + d$. Therefore $B = C + D$.

Assume $x \in C \cap D$. Since $x \in D$, there exists $b' \in B'$ such that $b' = a + x$, $a \in A$. Since $x \in C \cap A'$ and $A \subseteq A'$, we have $a + x \in A'$. Hence $b' \in B' \cap A' = 0$, and so $b' = 0$ and $a = x = 0$. Thus $B = C \oplus D$.

It follows from (ii) that

$$V = A \oplus C \oplus D = A' \oplus D. \tag{9}$$

Define now a map $\theta : V \longrightarrow V$ as follows:

$$\forall \quad a' \in A', \quad b' \in B' : (a' + b')^{\theta} = a' + b'^{\pi}. \tag{10}$$

An easy verification shows that $\theta \in \mathrm{Aut}\, V$.

(iii) $\theta \in \Phi \dot{\cap} \Phi'$. It suffices to show that θ acts identically on A and V/A. The first follows directly from (10). Further, take any $v \in V$, then $v = a' + b'$, where $a' \in A'$, $b' \in B'$. Let $b' = a + b$, where $a \in A$, $b \in B$. By (8) and (10), $v^{\theta} = a' + b$, whence $v - v^{\theta} = a' + b' - a' - b = a$, i.e. $v^{\theta} \equiv v \pmod A$, as required.

By (iii), θ is an element of G and so we can write $v \circ \theta$ as well as v^{θ}.

Denote by H the set of all $g \in G$ which act identically on A and B', and such that $C \circ g = C$. Since G_1' is the set of all $g \in G$ which act identically on B', if follows that $H \subseteq G_1'$.

(iv) $G = \langle \Psi, G_1, H, G_2' \rangle$. Since $G = \Phi' \rtimes (G_1' \times G_2')$ and $\Phi' \subseteq \Psi$, it suffices to prove that $G_1' \subseteq \langle \Psi, G_1, H, G_2' \rangle$. Denote the latter subgroup by G_0. Let $x \in G_1'$, then

$$x = \varphi g_1 g_2 \quad \text{for some} \quad \varphi \in \Phi, \quad g_i \in G_i.$$

Since $\Phi, G_1 \subseteq G_0$, it remains to show that $g_2 \in G_0$.

Take any $c \in C$. Recall that $C \subseteq A'$ and A' is G-invariant; therefore $c \circ g_2 \in \in A' = A + C$. Hence

$$\forall \quad c \in C : c \circ g_2 = a_c + c', \quad \text{where} \quad a_c \in A, \quad c' \in C. \tag{11}$$

Let $\tau = \tau(g_2)$ be an automorphism of the space V which acts trivially on A and B', but on C acts by the rule

$$\forall \quad c \in C : c \circ \tau = a_c + c. \tag{12}$$

Clearly τ stabilizes (6), i.e. $\tau \in \Psi$. Let us prove that $\tau^{-1} g_2 \in H$.

First, $\tau^{-1} g_2$ acts trivially on A by the definition. Next, if $c \in C$, then from (11) and (12)

$$c \circ (\tau^{-1} g_2) = (-a_c + c) \circ g_2 = -a_c + a_c + c' = c',$$

i.e. C is invariant under $\tau^{-1} g_2$. Finally, let $b' \in B'$. Since $x \in G_1'$,

$$b' \circ x = b'. \tag{13}$$

On the other hand, b' can be written in a form $b' = a + b$ $(a \in A,\ b \in B)$, and so

$$b' \circ x = (a + b) \circ \varphi g_1 g_2 = a \circ g_1 + a_1 \circ g_1 + b \circ g_2 \tag{14}$$

where $a_1 = b\varphi \in A$. Comparing (13) and (14), we obtain $b \circ g_2 = b$, whence

$$b' \circ g_2 = (a + b) \circ g_2 = a \circ g_2 + b \circ g_2 = a + b = b'.$$

Therefore g_2 acts identically on B', and so does $\tau^{-1} g_2$.

Combining all these observations, we conclude that $\tau^{-1} g_2 \in H$. But $\tau \in \Psi$, whence $g_2 \in G_o$. This completes the proof of (iv).

Consider the inner automorphism of (V, G) induced by θ^{-1}. By (10), θ acts identically on A, so that $(A, G_1)^{\theta^{-1}} = (A, G_1^{\theta^{-1}})$.

(v) $G_1^{\theta^{-1}} \subseteq G_1'$. For let $b' \in B'$, $\theta g \theta^{-1} \in G_1^{\theta^{-1}}$, where $g \in G_1$. We can write $b' = a + b$, where $a \in A$, $b \in B$, and then

$$b' \circ \theta g \theta^{-1} = (b' \circ \theta) \circ g) \circ \theta^{-1} = (b \circ g) \circ \theta^{-1} = b \circ \theta^{-1} = b'.$$

Thus $G_1^{\theta^{-1}}$ acts identically on B', whence (v) follows.

(vi) $(V, G) = (A, G_1^{\theta^{-1}}) \triangledown (C, H) \triangledown (B', G_2)$, $(A', G_1') = (A, G_1^{\theta^{-1}}) \triangledown (C, H)$.

1) H acts identically on A and B', but C is invariant under H. G_2' acts identically on $A' = A \oplus C$ and leaves B' invariant. G_1 acts identically on $B \supseteq C$, and θ acts identically on $A' \supseteq C$; therefore G_1^{θ} acts identically on C. Combining all these observations, (v) and (7), we obtain:

$$(V, \langle G_1^{\theta^{-1}}, H, G_2' \rangle) = (A, G_1^{\theta^{-1}}) \times (C, H) \times (B', G_2').$$

2) The normal subgroup Ψ of G is the stabilizer of the series $0 \subsetneq A \subseteq A \oplus C \subsetneq V$ in Aut V.

3) Clearly $\Psi \cap (G_1^{\theta^{-1}} \times H \times G_2') = 1$. From (iv) and $\theta \in \Psi$ we have $\langle \Psi, G_1^{\theta^{-1}}, H, G_2' \rangle = G$. Therefore $G = \Psi \rtimes (G_1^{\theta^{-1}} \times H \times G_2')$.

It follows from 1) - 3) that

$$(V, G) = (A, G_1^{\theta^{-1}}) \triangledown (C, H) \triangledown (B', G_2'). \tag{15}$$

Recall now that $(V, G) = (A', G_1') \triangledown (B', G_2')$, $(A, G_1^{\theta^{-1}}) \subseteq (A', G_1')$, and $(C, H) \subseteq$

$\subseteq (A', G_1')$. Taking all of this into account, it is easy to deduce that $(A', G_1') = (A, G_1^{\theta^{-1}}) \triangledown (C, H)$.

(vii) $G_2'^{\theta} \subseteq G_2$. Choose any $g \in G_2'^{\theta}$ and write it in the form $g = \theta^{-1} g_2' \theta$, where $g_2' \in G_2'$. To prove that $g \in G_2$, we must show that g acts identically on A and leaves B invariant. The first is evident since both G_2' and θ act identically on A'. Let $b \in B$, then according to (ii), $b = c + d$ for some $c \in C$, $d \in D$. But $d = b' - a$ for some $b' \in B'$, $a \in A$, so that $b = c + b' - a = (-a + c) + b'$. Since $-a + c \in A'$,

$$b \circ g = (-a + c) \circ g + b' \circ g = -a + c + b' \circ g,$$

On the other hand, since $b' \circ \theta^{-1} = a + b'$, we have

$$b' \circ g = b' \circ (\theta^{-1} g_2' \theta) = (a + b') \circ (g_2' \theta) = a + b' \circ (g_2' \theta).$$

Consequently

$$b \circ g = -a + c + b' \circ g = -a + c + a + b' \circ (g_2' \theta) = c + b \circ (g_2' \theta) \in B.$$

(viii) $(V, G) = (A, G_1) \triangledown (C, H) \triangledown (B', G_2')^{\theta}$, $(B, G_2) = (C, H) \triangledown (B', G_2')^{\theta}$.

Recal first that $B'^{\theta} = D$ and $V = A \oplus C \oplus D$. G_1 acts identically on $B = C \oplus D$. H acts identically on A and B', and since each $d \in D$ is presented in the form $d = b' - a$ for some $b' \in B'$, $a \in A$, H acts identically on $A \oplus D$. Finally, G_2' and θ act trivially on $A' = A \oplus C$. Therefore

$$(V, \langle G_1, H, G_2'^{\theta} \rangle) = (A, G_1) \times (C, H) \times (B', G_2')^{\theta}.$$

According to (i), Ψ, the stabilizer of (6), is a normal subgroup of G. The equality $G = \Psi \rtimes (G_1 \times H \times G_2')$ is proved now in the same manner as its analogue in (vi). Thus

$$(V, G) = (A, G_1) \triangledown (C, H) \triangledown (B', G_2')^{\theta}. \tag{16}$$

On the other hand $(V, G) = (A, G_1) \triangledown (B, G_2)$, $(C, H) \subseteq (B, G_2)$, and $(B', G_2')^{\theta} \subseteq (B, G_2)$ - see (vii). This shows that $(B, G_2) = (C, H) \triangledown (B', G_2')^{\theta}$, as claimed.

We have from (vi) and (viii) that (15) and (16) are refinements of (5) and (4) respectively. Since θ acts identically on A, C and H, (15) is conjugated with (16) by means of θ. This completes the first stage of the proof.

Stage 2. Now consider the general case. Let there are given arbitrary two triangular decompositions of $\rho = (V, G)$:

$$\rho = \rho_1 \nabla \rho_2 \nabla \dots \nabla \rho_m, \tag{17}$$

$$\rho = \rho_1' \nabla \rho_2' \nabla \dots \nabla \rho_n'. \tag{18}$$

If $m = 1$ or $n = 1$, Theorem 5.1. is trivial. Therefore we shall assume that $m, n \quad 1$. We proceed by induction on $m + n$. The case $m + n = 4$ has been proved in Stage 1. Suppose now that for $m + n = r$ the Theorem has been already established, and let for (17) and (18) we have $m + n = r + 1$.

Denote $\sigma = \rho_2 \nabla \dots \nabla \rho_m$, $\sigma' = \rho_2' \nabla \dots \nabla \rho_n'$; then

$$\rho = \rho_1 \nabla \sigma = \rho_1' \nabla \sigma'.$$

If $\rho_1 = (A, G_1)$ and $\rho_1' = (A', G_1')$ then, as before, either $A \subseteq A'$ or $A' \subseteq A$. Let $A \subseteq A'$. As has been established in Stage 1, there exists a subrepresentation $\tau = (C, H)$ of ρ and an element θ, stabilizing the series $0 \subseteq A \subseteq A' \subseteq V$, for which

$$\rho = \rho_1 \nabla \tau \nabla \sigma'^{\theta} = \rho_1^{\theta^{-1}} \nabla \tau \nabla \sigma'; \tag{19}$$

$$\sigma = \tau \nabla \sigma'^{\theta}, \quad \rho_1' = \rho_1^{\theta^{-1}} \nabla \tau. \tag{20}$$

Moreover, both the decompositions in (19) are conjugated by 0.

By (20),

$$\rho_2 \nabla \dots \nabla \rho_m = \tau \nabla \rho_2'^{\theta} \nabla \dots \nabla \rho_n'^{\theta}. \tag{21}$$

The full number of factors in (21) equals $(m - 1) + 1 + (n - 1) = r$, whence by induction hypothesis the decompositions (21) have respectively refinements

$$\omega_1 \nabla \dots \nabla \omega_s \quad \text{and} \quad \omega_1' \nabla \dots \nabla \omega_s' \tag{22}$$

conjugated by some $\eta : \omega_i^{\eta} = \omega_i'$.

Since $\omega_1' \nabla \dots \nabla \omega_s'$ is a refinement of $\tau \nabla \rho_2'^{\theta} \nabla \dots \nabla \rho_n'^{\theta}$, it follows that $\omega_1'^{\theta^{-1}} \nabla \dots \nabla \omega_s'^{\theta^{-1}}$ is a refinement of $\tau \nabla \rho_2' \nabla \dots \nabla \rho_n' = \tau \nabla \sigma'$. This together with (19) and (20) implies

$$\rho = \rho_1^{\theta^{-1}} \nabla \omega_1'^{\theta^{-1}} \nabla \dots \nabla \omega_s'^{\theta^{-1}} \tag{23}$$

and, moreover, (23) is a refinement of (18).

Since $\omega_1 \triangledown \dots \triangledown \omega_s$ is a refinement of $\rho_2 \triangledown \dots \triangledown \rho_m$,

$$\rho = \rho_1 \triangledown \omega_1 \triangledown \dots \triangledown \omega_s \tag{24}$$

is a refinement of (17). Further, being an element of the acting group of $\rho_2 \triangledown \dots \triangledown \rho_m$, η acts trivially on ρ_1. Hence we can rewrite (23) as follows:

$$\rho = \rho_1^{\eta\theta^{-1}} \triangledown \omega_1^{\eta\theta^{-1}} \triangledown \dots \triangledown \omega_s^{\eta\theta^{-1}}. \tag{25}$$

Thus (17) and (18) have refinements (24) and (25), respectively, conjugated by $\eta\theta^{-1}$. ∎

Thus, starting from the theorem of P.Neuman cited above, we have received a statement of quite different nature. The established theorem (but not its proof) rather reminds of the classical Krull-Remak-Schmidt Theorem asserting that if a group G possesses a chief series, then any two direct decompositions of G have central isomorphic refinements (see, for example, [33] , [62] and [35 ; §§ 47, 47a]).In our case, the pivot of a triangular product plays the role of the chief series, but the conjugation by means of $\eta\theta^{-1}$ is somewhat similar to the central isomorphism.

There naturally arises a question whether Theorem 5.1 can be generalized to arbitrary triangular products in the sense of Section 4. We believe that the answer is in the affirmative.

5.3. PROBLEM. To generalize Theorem 5.1 to triangular products of arbitrary partially ordered collections of representations.

We will make now a few remarks on the group of automorphisms of a triangular product. For a representation $\rho = (V, G)$, the group of all automorphisms of ρ is denoted by $\mathrm{Aut}\,\rho = \mathrm{Aut}\,(V, G)$. A G-submodule of V is called <u>characteristic</u> in ρ if it is invariant under all automorphisms of ρ. For example, if $(V, G) = (A, G_1) \triangledown (B, G_2)$ with finite-dimensional A then, by 2.10, A is characteristic in (V, G).

Now let (A, G_1) and (B, G_2) be two faithful representations. If $H_1 = \mathrm{Aut}\,(A, G_1)$ and $H_2 = \mathrm{Aut}\,(B, G_2)$, then there naturally arise representations (A, H_1) and

(B, H_2). Consider two triangular products:

$$(V, G) = (A, G_1) \triangledown (B, G_2) \quad \text{and} \quad (V, H) = (A, H_1) \triangledown (B, H_2).$$

5.4. PROPOSITION (Bunt [6]). If A is characteristic in (V, G), then $\mathrm{Aut}\,(V, G) = H$.

For the proof see [6] . Note that, in general, the submodule A need not be characteristic in $(A, G_1) \triangledown (B, G_2)$ as the following example shows. Let ρ be any representation and $\mathbb{Z}$ the set of integers in their natural order. For each $n \in \mathbb{Z}$ denote by ρ_n an isomorphic copy of ρ, and let

$$\sigma = \mathop{\triangledown}_{n \in \mathbb{Z}} \rho_n.$$

In view of the associativity of triangular products, we have

$$\sigma = (A, G_1) \triangledown (B, G_2)$$

where $(A, G_1) = \mathop{\triangledown}\limits_{n=-\infty}^{-1} \rho_n$, $(B, G_2) = \mathop{\triangledown}\limits_{n=0}^{+\infty} \rho_n$. Consider a natural automorphism of σ mapping ρ_n onto ρ_{n+1} for every $n \in \mathbb{Z}$. It is evident that A is not invariant under this automorphism.

We remark, in conclusion, that the transition from a representation (V, G) to the representation $(V, \mathrm{Aut}\,(V, G))$, noted above, leads to an interesting question. Let $\rho = (V, G)$ be a <u>faithful</u> representation over an arbitrary ring and let $G_1 = \mathrm{Aut}\,\rho$. It is easy to see that the representation $\rho_1 = (V, G_1)$ is also faithful, and that the map $(V, G) \longrightarrow (V, G_1)$, identical on V and assigning to each $g \in G$ the corresponding inner automorphism of ρ, is a right monomorphism. Therefore G_1 may (and will) be regarded as a subgroup of $\mathrm{Aut}\,V$ containing G, but ρ as a subrepresentation of ρ_1 (moreover, it is not hard to understand that $G_1 = N_{\mathrm{Aut}\,V}(G)$, the normalizer of G in $\mathrm{Aut}\,V$). Denote $\rho_1 = \alpha(\rho)$ and then define inductively $\rho_{\lambda+1} = \alpha(\rho_\lambda)$, $\rho_\mu = \bigcup_{\lambda<\mu} \rho_\lambda$ for ordinals λ and limit ordinals μ. We obtain an ascending series of faithful representations

$$\rho = \rho_0 \subseteq \rho_1 \subseteq \dots \rho_\lambda \subseteq \rho_{\lambda+1} \subseteq \dots$$

which will be called the <u>automorphism tower</u> of ρ (by analogy with a well known group theoretic notion going back to Wielandt [80]). Since the acting groups of all the ρ_λ are subgroups of Aut V, this tower has a definite height - the least ordinal $\eta = \eta(\rho)$ such that $\rho_\eta = \rho_{\eta+1} = \dots$. There naturally arises the general question of calculating this height for a given faithful representation.

Of course, the question is too general, and we should restrict ourselves to certain "good subquestions" of it. For example:

5.5. PROBLEM. Let ρ be a faithful finite-dimensional representation over a field. Is the automorphism height of ρ finite ?

EQUIVALENTLY: Let K be a field, $G = GL_n(K)$ and H a subgroup of G. Does the series

$$H \subseteq N_G(H) \subseteq N_G(N_G(H)) \subseteq \dots$$

terminate after a finite number of steps ?

This question is inspired by a well known theorem of Wielandt [80] on automorphism towers of finite groups.

6. Identities of triangular products

The material of this section shows the role of triangular products in the investigation of the semigroup of varieties of group representations.

Before stating the main result, consider arbitrary representations (A, G_1) and (B, G_2) and set $(V, G) = (A, G_1) \triangledown (B, G_2)$. Our first objective consists in the following: for a given word $u(x_1, \dots, x_n) \in KF$ and given $g_1, \dots, g_n \in G$, to represent the element $u(g_1 \dots, g_n) \in KG$ in a certain convenient form (for the notation see Section 1).

First let us calculate $f(g_1, \dots, g_n)$ for a group word $f(x_1, \dots, x_n) \in F$. By the definition of triangular product,

$$G = \mathrm{Hom}\,(B, A) \leftthreetimes (G_1 \times G_2)$$

where the group Hom (B, A) is written multiplicatively. Now it will be convenient to denote by Φ the group Hom (B, A) with the usual additive notation, but by $\overline{\Phi}$ the same group written multiplicatively, the symbol $\overline{}$ denoting the natural isomorphism $\Phi \longrightarrow \overline{\Phi}$. Using such a notation, we have, in particular,

$$b \circ \overline{\varphi} = b + b\varphi \quad (b \in B, \varphi \in \Phi).$$

Let $g_i = \overline{\varphi}_i h_i$, where $\varphi_i \in \Phi$, $h_i \in G_1 \times G_2$. Then

$$f(g_1,\dots,g_n) = f(\overline{\varphi}_1 h_1,\dots,\overline{\varphi}_n h_n) = \overline{\varphi}_1^{u'_1} \dots \overline{\varphi}_n^{u'_n} f(h_1,\dots,h_n) =$$
$$= \overline{(\varphi_1 \circ u'_1 + \dots + \varphi_n \circ u'_n)}\; f(h_1,\dots,h_n).$$

Here $u_i = u_i(x_1,\dots x_n)$ are certain elements of the group algebra $\mathbb{Z}F$ which depend only on the word $f(x_1,\dots,x_n)$ (in fact, u_i are the so-called Fox derivatives of $f(x_1, \dots,x_n)$ - see [12]), but $u'_i = u_i(h_1,\dots,h_n)$. Furthermore, one should remember that $G_1 \times G_2$ acts on Φ, and so does the group algebra $\mathbb{Z}[G_1 \times G_2]$.

If $v = a + b \in V = A \oplus B$, then

$$v \circ f(g_1,\dots,g_n) = (a + b(\varphi_1 \circ u'_1 + \dots + \varphi_n \circ u'_n) + b) \circ f(h_1,\dots,h_n) =$$
$$= v \circ f(h_1,\dots,h_n) + b(\varphi_1 \circ u'_1 + \dots + \varphi_n \circ u'_n) \circ f(h_1,\dots,h_n).$$

Now let $u(x_1,\dots,x_n) = \sum_k \lambda_k f_k(x_1,\dots,x_n)$, where $\lambda_k \in K$, $f_k(x_1,\dots,x_n) \in F$.

Then

$$u(g_1,\dots,g_n) = \sum \lambda_k f_k(g_1,\dots,g_n) =$$
$$= \sum \lambda_k \overline{(\varphi_1 \circ u'_{1k} + \dots + \varphi_n \circ u'_{nk})} f_k(h_1,\dots,h_n),$$

and so

$$v \circ u(g_1,\dots,g_n) = \sum \lambda_k (v \circ f_k(g_1,\dots,g_n)) =$$
$$= \sum \lambda_k (v \circ f_k(h_1,\dots,h_n) + b(\varphi_1 \circ u'_{1k} + \dots + \varphi_n \circ u'_{nk}) \circ f_k(h_1,\dots,h_n)) =$$
$$= v \circ u(h_1,\dots,h_n) + \sum_k \lambda_k b(\sum_i \varphi_i u'_{ik}) \circ f_k(h_1,\dots,h_n).$$

Suppose $u(x_1,\dots,x_n)$ is an identity of the representations (A, G_1) and (B, G_2). The same holds in their direct product $(A \oplus B, G_1 \times G_2)$, therefore $v \circ u(h_1,\dots,h_n) = 0$ and

$$v \circ u(g_1,\dots,g_n) = \sum_{k,i} \lambda_k b(\varphi_i \circ u'_{ik}) \circ f_k(h_1,\dots,h_n).$$

If $h_i = \overline{g}_i \overline{\overline{g}}_i$, where $\overline{g}_i \in G_1$, $\overline{\overline{g}}_i \in G_2$, then $f_k(h_1,\dots,h_n) = f_k(\overline{g}_1,\dots,\overline{g}_n)\, f_k(\overline{\overline{g}}_1,\dots,\overline{\overline{g}}_n)$.

Consequently, having in mind that $b(\varphi_i \circ u'_{ik}) \in A$, we obtain

$$v \circ u(g_1, \ldots, g_n) = \sum_{k,i} \lambda_k b(\varphi_i \circ u'_{ik}) \circ f_k(h_1, \ldots, h_n).$$

To obtain the formula we are interested in, it remains to decipher the elements $b(\varphi_i \circ u'_{ik})$. Since $u_{ik} = u_{ik}(x_1, \ldots, x_n) \in \mathbb{Z}F$,

$$u_{ik} = \sum_t n_{ikt} s_{ikt}$$

where $n_{ikt} \in \mathbb{Z}$, $s_{ikt} = s_{ikt}(x_1, \ldots, x_n) \in F$. Denote $s_{ikt}(\bar{g}_1, \ldots, \bar{g}_n) = \bar{s}_{ikt}$, $s_{ikt}(\bar{\bar{g}}_1, \ldots, \bar{\bar{g}}_n) = \bar{\bar{s}}_{ikt}$. Then

$$u_{ik}(h_1, \ldots, h_n) = \sum_t n_{ikt} \bar{s}_{ikt} \bar{\bar{s}}_{ikt}$$

and, furthermore,

$$b(\varphi_i \circ u'_{ik}) = b(\varphi_i \circ u_{ik}(h_1, \ldots, h_n)) = \sum_t n_{ikt} b(\varphi_i \circ \bar{s}_{ikt} \bar{\bar{s}}_{ikt}) =$$
$$= \sum_t n_{ikt}((b \circ \bar{\bar{s}}_{ikt}^{-1}) \varphi_i) \circ \bar{s}_{ikt}.$$

This yields the formula we need (Plotkin [54]):

$$v \circ u(g_1, \ldots, g_n) = \sum_{k,i,t} \lambda_k n_{ikt}(((b \circ \bar{\bar{s}}_{ikt}^{-1}) \varphi_i) \circ \bar{s}_{ikt}) \circ f_k(\bar{g}_1, \ldots, \bar{g}_n). \qquad (*)$$

We emphasize that in (*) <u>the integers n_{ikt} and the group words $s_{ikt}(x, \ldots, x_n)$ depend only on the initial word $u(x_1, \ldots, x_n)$ and that it holds only when $u(x_1, \ldots, x_n)$ is an identity of (A, G_1) and (B, G_2).</u>

Suppose now that we know all the identities of representations ϱ_1 and ϱ_2. What can we say about the identities of $\varrho_1 \triangledown \varrho_2$? Equivalently: how to calculate var $(\varrho_1 \triangledown \triangledown \varrho_2)$ provided we know var ϱ_1 and var ϱ_2 ? If the basic ring is a field, this question has an exhaustive answer.

6.1.THEOREM (Plotkin [58, 52], Vovsi [71]). If ϱ_1 and ϱ_2 are arbitrary representations over a field, then

$$\text{var}(\varrho_1 \triangledown \varrho_2) = \text{var}\,\varrho_1 \cdot \text{var}\,\varrho_2.$$

This theorem was established in the following way. First, Plotkin [58, 52] proved that if ϱ_1 and ϱ_2 are representations over a field and $\varrho_2^{(I)}$ is a countable direct power of ϱ_2, then var $(\varrho_1 \triangledown \varrho_2^{(I)}) = \text{var}\,\varrho_1 \cdot \text{var}\,\varrho_2$. Then the author [71] showed that the "exponent" I in this formula can be removed.

In fact, we will prove not the above theorem but a generalization of it which relates to representations over an arbitrary integral domain. Beforehand, let us give two definitions.

Let $\rho = (A, G)$ be a representation over K. If A is a projective K-module we shall say that ρ is a projective representation (although this term is far from being good)*. A variety $\mathfrak{X}$ is called projective if $\mathrm{Fr}\,\mathfrak{X}$ is a projective representation.

6.2. THEOREM (Vovsi [78]). If ρ_1 and ρ_2 are projective representations over an integral domain and var ρ_2 is a projective variety, then

$$\mathrm{var}\,(\rho_1 \triangledown \rho_2) = \mathrm{var}\,\rho_1 \cdot \mathrm{var}\,\rho_2.$$

Clearly this theorem contains the previous one. Its proof, incorporating methods of [54], [71] and some other techniques, divides into several lemmas.

6.3. LEMMA [31]. Let $\mathfrak{X}$ be an arbitrary variety and $\mathfrak{Y}$ a projective variety. Then

$$\mathrm{var}\,(\mathrm{Fr}\,\mathfrak{X} \triangledown \mathrm{Fr}\,\mathfrak{Y}) = \mathfrak{X}\mathfrak{Y}.$$

P r o o f. Let $I = \mathrm{Id}\,\mathfrak{X}$, $J = \mathrm{Id}\,\mathfrak{Y}$. Then $JI = \mathrm{Id}(\mathfrak{X}\mathfrak{Y})$ and

$$(KF/JI, F) = \mathrm{Fr}(\mathfrak{X}\mathfrak{Y}).$$

Denote $KF/JI = \overline{KF}$, $J/JI = \overline{J}$, then

$$(\overline{KF}/\overline{J}, F) \cong (KF/J, F) = \mathrm{Fr}\,\mathfrak{Y}$$

and, by hypothesis, $\overline{KF}/J$ is a projective K-module. Hence $\overline{J}$ has a direct complement in $\overline{KF}$.

Let $(\overline{KF}, G)$, $(\overline{J}, G_1)$, and $(\overline{KF}/\overline{J}, G_2)$ be the faithful images of $(\overline{KF}, F)$, $(\overline{J}, F)$ and $(\overline{KF}/\overline{J}, F)$ respectively. By the Embedding Theorem,

$$(\overline{KF}, G) \hookrightarrow (\overline{J}, G_1) \triangledown (\overline{KF}/\overline{J}, G_2),$$

---

* In this case A is also called a KG-lattice

and since the latter belongs to $\mathfrak{X}\mathfrak{Y}$,

$$\text{var}\ ((\bar{J}, G_1) \triangledown (\overline{KF}/\bar{J}, G_2)) = \mathfrak{X}\mathfrak{Y}. \tag{1}$$

Denote, for brevity, $KF/I = E_1$ and $KF/J = E_2$. Since $(\bar{J}, F) \in \mathfrak{X}$, $(\bar{J}, F)$ is an epimorphic image of some free representation (E, F) of $\mathfrak{X}$:

$$\mu : (E, F) \longrightarrow (\bar{J}, F).$$

Here E is a direct power of the KF-module E_1, but μ acts trivially on F.

By hypothesis, E_2 is a projective K-module. Therefore, by Proposition 2.6, there is an epimorphism

$$\bar{\mu} : (E, F) \triangledown (E_2, F) \longrightarrow (\bar{J}, F) \triangledown (E_2, F). \tag{2}$$

Since the faithful image of $(\bar{J}, F) \triangledown (E_2, F)$ is $(\bar{J}, G_1) \triangledown (E_2, G_2)$, it follows from (1) that $\text{var}\ ((\bar{J}, F) \triangledown (E_2, F)) = \mathfrak{X}\mathfrak{Y}$. By (2),

$$\text{var}\ ((E, F) \triangledown (E_2, F)) = \mathfrak{X}\mathfrak{Y}.$$

Since E, as a KF-module, is a direct power of E_1, the representation $(E, F) \triangledown (E_2, F)$ is a "residually $(E_1, F) \triangledown (E_2, F)$" (see Lemma 3.4). Hence $\text{var}\ ((E_1, F) \triangledown (E_2, F)) = \mathfrak{X}\mathfrak{Y}$. ■

6.4. COROLLARY. Let $\mathfrak{X} = \text{var}\ (A, G)$ and let $\mathfrak{Y}$ be a projective variety. Then

$$\text{var}\ ((A, G) \triangledown \text{Fr}\,\mathfrak{Y}) = \mathfrak{X}\mathfrak{Y}.$$

P r o o f. By the preceding lemma,

$$\mathfrak{X}\mathfrak{Y} = \text{var}\ (\text{Fr}\,\mathfrak{X} \triangledown \text{Fr}\,\mathfrak{Y}). \tag{3}$$

Proposition 1.8 shows that $\text{Fr}\,\mathfrak{X} \in \text{VSC}\{(A, G)\}$, whence, by Lemma 3.4,

$$\text{Fr}\,\mathfrak{X} \triangledown \text{Fr}\,\mathfrak{Y} \in (\text{VSC}\{(A, G)\}) \triangledown \text{Fr}\,\mathfrak{Y} \subseteq \text{VSA}\{(A, G) \triangledown \text{Fr}\,\mathfrak{Y}\} \subseteq$$

$$\subseteq \text{var}\ ((A, G) \triangledown \text{Fr}\,\mathfrak{Y}).$$

From this and (3) we obtain the necessary equality. ■

From now on in this section, K is an arbitrary integral domain and P its field of fractions. All modules and representations as well as functors Hom and $\otimes$ are considered over K.

For any K-module A the K-module $A_P = P \otimes_K A$ will be called the P-envelope of A. It is well known that if A is a torsion-free module, it can be naturally embedded in A_P. The dimension of the vector P-space A_P is called the rank of A. If $\varrho = (A, G)$ is any representation over K, then the group G naturally acts on A_P by the rule

$$(\lambda \otimes a) \circ g = \lambda \otimes (a \circ g).$$

Thus there arises a representation $\varrho_P = (A_P, G)$ which will be called the P-envelope of the representation ϱ.

6.5. LEMMA. Let (B, G) be a representation with torsion-free B and let (E, F) be a free representation of var (B, G). Then for each finite set $a_1, \dots, a_n \in E$ there exists a finite direct power $(B, G)^m$ and a homomorphism of P-envelopes

$$\nu : (E_P, F) \to (B_P, G)^m$$

which is injective on the submodule $\langle a_1, \dots, a_n \rangle_P$.

P r o o f. Note first that, by Proposition 1.8,

$$(E, F) \in \mathrm{VSC}\{(B, G)\}. \tag{4}$$

Hence E is a submodule of a Cartesian power of B and so E is torsion-free. Consequently E can be embedded into E_P and $A = \langle a_1, \dots, a_n \rangle$ can be regarded as a submodule of E_P; this is assumed in the Lemma.

It follows from (4) that there exists a Cartesian power $(B, G)^I$ and a homomorphism

$$\mu : (E, F) \to (B, G)^I$$

which is injective on E. It induces a homomorphism

$$\mu_P : (E_P, F) \to (B_P, G)^I$$

of envelopes which is injective on E_P. Since A_P is a finite-dimensional P-subspace of E_P, $C = A_P^{\mu_P}$ is a subspace of the same finite dimension in B_P^I.

Let M be an arbitrary finite subset of I, $\pi_M : (B_P, G)^I \to (B_P, G)^M$ the natural projection, and D_M the kernel of the map $\pi_M : B^I \to B^M$. Since the intersection of all D_M over all finite subsets M is zero, the intersection of all $C \cap D_M$ is zero as well. Since C is finite-dimensional and since $D_M \cap D_N = D_{M \cup N}$, it is easy to understand that $C \cap D_M = 0$ for some finite M. This means that the corresponding projection π_M is injective on C. Therefore the composition of homomorphisms

$$\mu_P \pi_M : (E_P, F) \longrightarrow (B_P, G)^M$$

is injective on A_P. It remains to set $\nu = \mu_P \pi_M$.

6.6. LEMMA. Let (A, G_1) and (B, G_2) be representations such that the module A is free and the module B is projective. Then

$$(A_P, G_1) \triangledown (B_P, G_2) \in \mathrm{var}\,((A, G_1) \triangledown (B, G_2)).$$

P r o o f. Choose in the group algebra KF an arbitrary element

$$u(x_1, \ldots, x_n) = \sum \lambda_k f_k(x_1, \ldots, x_n),$$

where $\lambda_k \in K$, $f_k(x_1, \ldots, x_n) \in F$. It suffices to prove that if the bi-identity $y \circ u(x_1, \ldots$ $\ldots, x_n) = 0$ is not satisfied in $(A_P, G_1) \triangledown (B_P, G_2)$, then it is not satisfied in $(A, G_1) \triangledown (B, G_2)$ either. If it is not satisfied in (A, G_1), then it is nothing to prove. So we may assume that $u(x_1, \ldots, x_n)$ is an identity of (A, G_1). Evidently the same holds in (A_P, G_1).

Thus let $y \circ u(x_1, \ldots, x_n) = 0$ fails in $(V, G) = (A_P, G_1) \triangledown (B_P, G_2)$ but is satisfied in (A_P, G_1). Then there exist elements $b \in B_P$, $g_1, \ldots, g_n \in G$ such that $b \circ u(g_1, \ldots, g_n) \neq 0$. Let $g_i = \varphi_i \bar{g}_i \bar{\bar{g}}_i$, where $\varphi_i \in \mathrm{Hom}\,(B_P, A)$, $\bar{g}_i \in G_1$, $\bar{\bar{g}}_i \in G_2$. By (*),

$$b \circ u(g_1, \ldots, g_n) = \sum_{k,i,t} \lambda_k n_{ikt} (((b \circ \bar{\bar{s}}_{ikt}^{-1}) \varphi_i) \circ \bar{s}_{ikt}) \circ f_k(\bar{g}_1, \ldots, \bar{g}_n), \tag{5}$$

where n_{ikt} are integers and $s_{ikt} = s_{ikt}(x_1, \ldots, x_n)$ are group words depending only on $u(x_1, \ldots, x_n)$, $\bar{s}_{ikt} = s_{ikt}(\bar{g}_1, \ldots, \bar{g}_n)$, $\bar{\bar{s}}_{ikt} = s_{ikt}(\bar{\bar{g}}_1, \ldots, \bar{\bar{g}}_n)$.

Since the module B is projective, it is a direct summand of some free module $\bar{B} : \bar{B} = B \oplus B'$. Having defined an action of G_2 on B' in the trivial way, we obtain a representation $(\bar{B}, G_2)$. Since

$$\bar{B}_P = P \otimes \bar{B} = (P \times B) \oplus (P \otimes B') = B_P \oplus B'_P$$

and

$$\mathrm{Hom}\,(\bar{B}_P, A_P) = \mathrm{Hom}\,(B_P, A_P) \oplus \mathrm{Hom}\,(B_P', A_P),$$

we can assume that the elements b and φ_i belong to $\bar{B}_P$ and $\mathrm{Hom}\,(\bar{B}_P, A_P)$, respectively. It follows, in view of (5), that $u(x_1,\dots,x_n)$ is not an identity of $(A_P, G_1)\triangledown$ $\triangledown(\bar{B}_P, G_2)$. Let us consider two cases.

<u>(i) The free module $\bar{B}$ has a finite rank.</u> Since the module A is also free, it is easy to understand that

$$\mathrm{Hom}\,(P\otimes\bar{B},\ P\otimes A) = P\otimes\mathrm{Hom}\,(\bar{B}, A). \tag{6}$$

Hence there exists $\alpha\in P$ such that every $\varphi_i (i = 1,\dots,n)$ can be written in a form

$$\varphi_i = \alpha\psi_i,\quad \text{where}\quad \psi_i\in\mathrm{Hom}\,(\bar{B}, A).$$

Denote by ξ_i the restriction of ψ_i to B and take elements $h_i = \xi_i g_i \bar{g}_i \bar{\bar{g}}_i$ in the acting group of $(A, G_1)\triangledown(B, G_2)$. Furthermore, $b = \beta c$ for some $\beta\in P$ and $c\in B$. Using (5), we have:

$$0\neq b\circ u(g_1,\dots,g_n) = \sum\lambda_k n_{ikt}(((\beta c\circ\bar{\bar{s}}_{ikt}^{-1})\alpha\psi_i)\circ\bar{s}_{ikt})\circ f_k(\bar{g}_1,\dots,\bar{g}_n) =$$
$$= \alpha\beta\sum\lambda_k n_{ikt}(((c\circ\bar{\bar{s}}_{ikt}^{-1})\psi_i)\circ\bar{s}_{ikt})\circ f_k(\bar{g}_1,\dots,\bar{g}_n) =$$
$$= \alpha\beta\,(c\circ u(h_1,\dots,h_n)).$$

Therefore the bi-identity $y\circ u(x_1,\dots,x_n) = 0$ fails in $(A, G_1)\triangledown(B, G_2)$, as required.

<u>(ii) B has an infinite rank.</u> Then the formula (6) fails and, in general, we cannot find such a fixed element $\alpha\in P$ that $\varphi_i = \alpha\psi_i$, $\psi_i\in\mathrm{Hom}\,(\bar{B}, A)$ for all i.

However, this difficulty can be overcome. Notice that the only essential property of the homomorphisms $\varphi_i\in\mathrm{Hom}\,(\bar{B}_P, A_P)$ is their action on the finite set of elements $\{b\circ\bar{\bar{s}}_{ikt}^{-1}\}$. Therefore these homomorphisms can be "corrected" in the following way. Pick a basis $\{e_i\}$ of the free K-module $\bar{B}$;it is also a basis of the vector space $\bar{B}_P$ over P. From this basis choose a finite set $e_1,\dots,e_r$ such that all $b\circ\bar{\bar{s}}_{ikt}^{-1}$ belong to its linear P-envelope $\langle e_1,\dots,e_r\rangle_P = C$. Let C' be the subspace generated by all the other elements of the basis $\{e_i\}$, then $\bar{B}_P = C\oplus C'$. Replace every homomorphism $\varphi_i\in$

Hom $(\bar{B}_p, A_p)$ by a homomorphism φ_i^o defined as follows:

$$x\varphi_i^o = x\varphi_i \qquad \text{if } x \in C,$$

$$x\varphi_i^o = 0 \qquad \text{if } x \in C'.$$

For these φ_i^o $(i = 1,\dots,n)$ we can find such a fixed $\alpha \in P$ that $\varphi_i^o = \alpha\psi_i$, $\psi_i \in \mathrm{Hom}\,(\bar{B}, A)$, and then the proof is continued as in (i). ■

6.7. LEMMA. Let (A, G_1) and (B, G_2) be representations such that the module A is free and B is projective. Denote $\mathfrak{X} = \mathrm{var}\,(A, G_1)$, $\mathfrak{Y} = \mathrm{var}\,(B, G_2)$ and let $\mathfrak{Y}$ be projective. Then

$$\mathrm{var}\,((A, G_1) \triangledown D\{(B, G_2)\}) = \mathfrak{X}\mathfrak{Y}.$$

P r o o f. As usual, $D\{(B, G_2)\}$ denotes the class of all direct powers of the representation (B, G_2). Let $(E, F) = \mathrm{Fr}\,\mathfrak{Y}$, then by Corollary 6.4,

$$\mathfrak{X}\mathfrak{Y} = \mathrm{var}\,((A, G_1) \triangledown (E, F)).$$

Hence it is enough to show that

$$(A, G_1) \triangledown (E, F) \in \mathrm{var}\,((A, G_1) \triangledown D\{(B, G_2)\}).$$

In other words, we have to prove that if $u(x_1,\dots,x_n) \in KF$ is not an identity of $(A, G_1) \triangledown (E, F)$, it is not an identity of some representation of the class $(A, G_1) \triangledown D\{(B, G_2)\}$. As in the proof of Lemma 6.6, we may assume $u(x_1,\dots,x_n)$ to be an identity of (A, G_1).

Since $y \circ u(x_1,\dots,x_n) = 0$ fails in

$$(V, G) = (A, G_1) \triangledown (E, F) = (A \oplus E,\ \mathrm{Hom}\,(E, A) \leftthreetimes (G_1 \times F)),$$

there exist $b \in E$, $g_1,\dots,g_n \in G$ such that

$$b \circ u(g_1,\dots,g_n) \neq 0.$$

Let $u(x_1,\dots,x_n) = \sum \lambda_k f_k(x_1,\dots,x_n)$ where $\lambda_k \in K$, $f_k(x_1,\dots,x_n) \in F$; $g_i = \varphi_i \bar{g}_i \bar{\bar{g}}_i$ where $\varphi_i \in \mathrm{Hom}\,(E, A)$, $\bar{g}_i \in G_1$, $\bar{\bar{g}}_i \in F$. By (*),

$$b \circ u(g_1,\dots,g_n) = \sum_{k,i,t} \lambda_k n_{ikt}(((b \circ \bar{\bar{s}}_{ikt}^{-1})\varphi_i) \circ \bar{s}_{ikt}) \circ f_k(\bar{g}_1,\dots,\bar{g}_n) \tag{8}$$

Denote by M the submodule of E generated by all $b \circ \bar{\bar{s}}_{ikt}^{-1}$. By Lemma 6.5, there exists such a homomorphism

$$\nu : (E_P, F) \longrightarrow (B_P, G_2)^m$$

(m is a positive integer) which is injective on M_P. Let $M_P^{\nu} = C$. C is a subspace of the vector space B_P^m, and so C is a direct summand of B_P^m. Therefore there exists a P-linear map $\eta: B_P^m \longrightarrow E_P$ such that the restriction of η to M_P is unity.

Suppose now that $u(x_1,\dots,x_n)$ is an identity of the class $(A, G_1)\nabla D\{(B, G_2)\}$. In particular, it is satisfied in $(A, G_1)\nabla(B, G_2)^m$ and, by Lemma 6.6, it is satisfied in

$$(A_P, G_1)\nabla(B_P, G_2)^m = (A_P \oplus B_P^m, \text{Hom}\,(B_P^m, A_P)\,\lambda(G_1 \times G_2^m)).$$

Since E and A are torsion-free, the elements $\varphi_i \in \text{Hom}\,(E, A)$ can be extended, in a natural way, to the homomorphisms $\bar{\varphi}_i \in \text{Hom}_K(E_P, A_P)$ of P-envelopes. Let $\psi_i = \eta\bar{\varphi}_i$, then $\psi_i \in \text{Hom}_K(B_P^m, A_P)$ and

$$(b\circ\bar{\bar{s}}_{ikt}^{-1})^{\nu}\psi_i = (b\circ\bar{\bar{s}}_{ikt}^{-1})^{\nu\eta}\bar{\varphi}_i = (b\circ\bar{\bar{s}}_{ikt}^{-1})\varphi_i, \tag{9}$$

(the latter equality is true since $\bar{\varphi}_i$ and φ_i agree on E).

Denote $b^{\nu} = c$, $\bar{\bar{g}}_i^{\nu} = h_i$, and $\tilde{s}_{ikt} = s_{ikt}(h_1,\dots,h_n)$; then $c \in B_P^m$, $h_i \in G_2^m$, $\tilde{s}_{ikt} \in G_2^m$, and

$$t_i = \psi_i\bar{g}_i h_i \in \text{Hom}\,(B_P^m, A_P)\,\lambda(G_1 \times G_2^m).$$

Consider the representation $(A_P, G_1)\nabla(B_P, G_2)^m$. Using (7), (8) and (9), we have:

$$\begin{aligned}
c\circ u(t_1,\dots,t_n) &= \sum\lambda_k n_{ikt}(((c\circ\tilde{s}_{ikt}^{-1})\psi_i)\circ\bar{s}_{ikt})\circ f_k(\bar{g}_1,\dots,\bar{g}_n) = \\
&= \sum\lambda_k n_{ikt}(((b^{\nu}\circ(\bar{\bar{s}}_{ikt}^{-1})^{\nu})\psi_i)\circ\bar{s}_{ikt})\circ f_k(\bar{g}_1,\dots,\bar{g}_n) = \\
&= \sum\lambda_k n_{ikt}(((b\circ\bar{\bar{s}}_{ikt}^{-1})^{\nu}\psi_i)\circ\bar{s}_{ikt})\circ f_k(\bar{g}_1,\dots,\bar{g}_n) = \\
&= \sum\lambda_k n_{ikt}(((b\circ\bar{\bar{s}}_{ikt}^{-1})\varphi_i)\circ\bar{s}_{ikt})\circ f_k(\bar{g}_1,\dots,\bar{g}_n) = \\
&= b\circ u(g_1,\dots,g_n) \neq 0.
\end{aligned}$$

Thus the bi-identity $y\circ u(x_1,\dots,x_n) = 0$ is not satisfied in $(A_P, G_1)\nabla(B_P, G_2)^m$. Contradiction. ∎

6.8. LEMMA. Let (A, G_1) and (B, G_2) be arbitrary projective representations and let $\bar{A}$ be a free module such that A is its direct summand: $\bar{A} = A \oplus A'$. Define an action of G_1 on A' in the trivial way. Then the arising representation

$(\bar{A}, G_1)$ satisfies the following conditions:

$$\operatorname{var}(A, G_1) = \operatorname{var}(\bar{A}, G_1),$$

$$\operatorname{var}((A, G_1) \triangledown (B, G_2)) = \operatorname{var}((\bar{A}, G_1) \triangledown (B, G_2)).$$

P r o o f. Since A is torsion-free, it is evident that $\operatorname{var}(A, G_1)$ contains the variety $\mathfrak{S}$ of trivial representations, whence $\operatorname{var}(A, G_1) = \operatorname{var}(\bar{A}, G_1)$. Denote

$$(V_1, T_1) = (A, G_1) \triangledown (B, G_2) = (A \oplus B, \operatorname{Hom}(B, A) \leftthreetimes (G_1 \times G_2)),$$

$$(V_2, T_2) = (\bar{A}, G_1) \triangledown (B, G_2) = (\bar{A} \oplus B, \operatorname{Hom}(B, \bar{A}) \leftthreetimes (G_1 \times G_2)),$$

and let $\mathfrak{X}_i = \operatorname{var}(V_i, T_i)$. Since the functor $\_\_\triangledown (B, G_2)$ is left exact, $(V_1, T_1) \hookrightarrow (V_2, T_2)$ and so $\mathfrak{X}_1 \subseteq \mathfrak{X}_2$.

To prove the reverse inclusion, observe that A and A' are T_2-submodules of V_2 and $A \cap A' = 0$. Hence it is enough to show that $(V_2/A, T_2) \in \mathfrak{X}_1$ and $(V_2/A', T_2) \in \mathfrak{X}_1$. First, remark that $(V_2/A', T_2) \sim (V_1, T_1)$, whence $(V_2/A', T_2) \in \mathfrak{X}_1$. Second, $(V_2/A, T_2) \sim (A', G_1) \triangledown (B, G_2)$. Since $(A', G_1) \in \mathfrak{S}$, we have $(A', G_1) \in \operatorname{var}(A, G_1) = VQSC\{(A, G_1)\}$, and so

$$(A', G_1) \triangledown (B, G_2) \in (VQSC\{(A, G_1)\}) \triangledown (B, G_2).$$

By Lemma 3.4, the operations V, S and C can be "carried out" from triangular products. Since B is projective, the operation Q can be carried out as well, in view of 2.6 (ii). Therefore

$$(A', G_1) \triangledown (B, G_2) \in VQSA\{(A, G_1) \triangledown (B, G_2)\} = \operatorname{var}(V_1, T_1) = \mathfrak{X}_1,$$

and so $(V_2/A, T_2) \in \mathfrak{X}_1$. ∎

P r o o f of T h e o r e m 6.2. Let ρ_1 and ρ_2 be projective representations, $\mathfrak{X} = \operatorname{var} \rho_1$, $\mathfrak{Y} = \operatorname{var} \rho_2$, and let $\mathfrak{Y}$ be projective. If $\rho_1 = (A, G_1)$, we may assume, in view of 6.8, that A is a free module. Using 6.7 and 3.4 we have

$$\mathfrak{X}\mathfrak{Y} = \operatorname{var}(\rho_1 \triangledown D\rho_2) \subseteq \operatorname{var}(RV(\rho_1 \triangledown \rho_2)) =$$

$$= \operatorname{var}(\rho_1 \triangledown \rho_2).$$

The reverse inclusion is obvious. ∎

Theorem 6.2 can be reformulated in a more general form. Namely, the representations ρ_1 and ρ_2 can be replaced by arbitrary classes of representations.

6.2.' THEOREM. Let $\mathfrak{K}_1$ and $\mathfrak{K}_2$ be arbitrary classes of projective representations and let var $\mathfrak{K}_2$ be projective. Then

$$\mathrm{var}\,(\mathfrak{K}_1 \triangledown \mathfrak{K}_2) = \mathrm{var}\,\mathfrak{K}_1 \cdot \mathrm{var}\,\mathfrak{K}_2.$$

P r o o f. For $i = 1, 2$ choose in the class $\mathfrak{K}_i$ a <u>set</u> of representations $\mathfrak{K}_i^o$ such that $\mathrm{var}\,\mathfrak{K}_i^o = \mathrm{var}\,\mathfrak{K}_i$. Let ρ_i be the direct product of all representations from $\mathfrak{K}_i^o$. Then ρ_i is projective, $\rho_i \in D\mathfrak{K}_i$ and $\mathrm{var}\,\mathfrak{K}_i = \mathrm{var}\,\rho_i$. Consequently, by 6.2 and 3.4,

$$\mathrm{var}\,\mathfrak{K}_1 \cdot \mathrm{var}\,\mathfrak{K}_2 = \mathrm{var}\,\rho_1 \cdot \mathrm{var}\,\rho_2 = \mathrm{var}\,(\rho_1 \triangledown \rho_2) \subseteq$$
$$\subseteq \mathrm{var}\,(D\mathfrak{K}_1 \triangledown D\mathfrak{K}_2) \subseteq \mathrm{var}\,(\mathfrak{K}_1 \triangledown \mathfrak{K}_2). \blacksquare$$

In conclusion, we remark a question naturally arising in connection with Theorem 6.2.

6.9. PROBLEM. Let ρ be a projective representation over an integral domain. Is the variety var ρ projective ?

If the answer is affirmative, the requirement "var ρ_2 is projective" in Theorem 6.2 is unnecessary. Suppose now that the answer is negative. It is not hard to deduce from 1.8 that the variety generated by a projective representation over an integral domain is <u>pure</u>, that is, its free representations have torsion-free domains of action. Thus we obtain an example of a pure variety which is not projective. This would solve Problem 15 of [55].

NOTE. If $K = \mathbb{Z}$, Problem 6.9 is solved in the affirmative (see [31], p.506).

Chapter 2

APPLICATIONS

7. Identities of triangular matrix groups and their canonical representations

This section is devoted to a very concrete problem. As usual, let K be a commutative ring with 1, and let $UT_n(K) = UT_n$ and $T_n(K) = T_n$ be the full unitriangular and triangular matrix groups of degree n over K, respectively. Denote by

$$ut_n = ut_n(K) = (K^n, UT_n(K)) \text{ and } t_n = t_n(K) = (K^n, T_n(K))$$

their canonical representations in the free K-module of rank n. These classical objects deserve to be studied from various positions; in particular, from the standpoint of identities and varieties. Naturally, the first problem one should solve here is the following: to describe the varieties var ut_n and var t_n or, equivalently, to find bases for the identities of the representations ut_n and t_n.

This question was solved in papers of Grinberg [13] and Kropí [30] by means of direct and sometimes rather long calculations in the group algebra KF. However, as has been pointed out by the author [71, 78], their results can be easily deduced from the results of Section 6. The corresponding proofs will be presented here.

7.1. THEOREM (Grinberg [13]). Over any ring

$$\text{var } ut_n = \mathfrak{S}^n.$$

P r o o f [71]. Take the representation (K, 1) where 1 is the unit group and K is the basic ring considered as a K-module. Evidently (K, 1) is the faithful image of the representation Fr $\mathfrak{S}$ = (KF/Δ, F). Hence $\mathfrak{S}$ is a projective variety and $\mathfrak{S}$ = = var (K, 1). By the definition of triangular product,

$$ut_n = \underbrace{(K, 1) \triangledown (K, 1) \triangledown \ldots \triangledown (K, 1)}_{n \text{ times}}.$$

Therefore, using Corollary 6.4, we obtain

$$\text{var } ut_2 = \text{var } ((K, 1) \triangledown (K, 1)) = \mathfrak{S}^2,$$

$$\text{var } ut_3 = \text{var } ((K, 1) \triangledown (K, 1)) \triangledown (K, 1)) = \mathfrak{S}^2 \cdot \mathfrak{S} = \mathfrak{S}^3,$$

- -

$$\text{var } ut_n = \text{var } ((\underbrace{(K, 1) \triangledown \ldots \triangledown (K, 1))}_{n-1} \triangledown (K, 1)) = \mathfrak{S}^{n-1} \cdot \mathfrak{S} = \mathfrak{S}^n. \blacksquare$$

7.2. COROLLARY. Over any ring the identity $(x_1 - 1)(x_2 - 1) \ldots (x_n - 1)$ forms a basis for the identities of the representation ut_n. ■

The problem of describing var t_n is more difficult. We will be able to reach such a description under the additional assumption that K is an integral domain.

Recall, following [52], that if $\mathfrak{X}$ is a variety of representations and Θ a variety of groups, then $\mathfrak{X} \times \Theta$ is the class of all representations (A, G) where G contains a normal subgroup H such that $(A, H) \in \mathfrak{X}$ and $G/H \in \Theta$. It is easy to see that $\mathfrak{X} \times \Theta$ is a variety of representations (for instance, by the Birkhoff Theorem), and that the following equality holds

$$(\mathfrak{X} \times \Theta) \cdot (\mathfrak{Y} \times \Theta) = (\mathfrak{X}\mathfrak{Y}) \times \Theta. \tag{1}$$

Consider a representation (K, K^*) where K^* is the multiplicative group of the ring K, acting on K by right multiplication. As usual, $\mathfrak{A}$ denotes the variety of all abelian groups, but $\mathfrak{A}_m$ the variety of abelian groups of exponent m.

7.3. LEMMA [30]. Let K be an integral domain. Then

$$\text{var } (K, K^*) = \begin{cases} \mathfrak{S} \times \mathfrak{A} & \text{if } |K^*| = \infty \\ \mathfrak{S} \times \mathfrak{A}_m & \text{if } |K^*| = m. \end{cases}$$

P r o o f in both cases is analogous, so we consider only the case $|K^*| = m$. First, it is clear that $(K, K^*) \in \mathfrak{S} \times \mathfrak{A}_m$. Next, let A be a free group of countable rank in $\mathfrak{A}_m$ with free generators $x_1, x_2, \ldots$, then $\overline{Fr}(\mathfrak{S} \times \mathfrak{A}_m) = \text{Reg } A = (KA, A)$. Hence it is enough to prove that $(KA, A) \in \text{var } (K, K^*)$. Consider all possible homomorphisms $\varphi_i : (KA, A) \longrightarrow (K, K^*)$. Let V_i and B_i be the kernels of φ_i in KA and A respectively, and let $V = \cap V_i$, $B = \cap B_i$. By the Remak Theorem,

$$(KA/V,\ A/B) \in SC\{(K, K^*)\},$$

whence $(KA/V, A) \in VSC\{(K, K^*)\} \subseteq \text{var}\ (K, K^*)$. Thus it is enough to show that $V = 0$.

Suppose that $0 \neq v(x_1,\dots,x_n) \in V$. This element can be uniquely presented as a polynomial over K in the variables $x_1,\dots,x_n$, reduced modulo the identity $x^m = 1$. A standard argument shows that there exist elements $\lambda_1,\dots,\lambda_n \in K^*$ such that $v(\lambda_1,\dots,\lambda_n) \neq 0$ (see, for example, Lang's "Algebra", Chapter 5, § 5). Consider now a homomorphism $\varphi_i : (KA, A) \longrightarrow (K, K^*)$ satisfying the condition $x_j^{\varphi_i} = \lambda_j$, $j = 1,\dots,n$. Then

$$(v(x_1,\dots,x_n))^{\varphi_i} = v(\lambda_1,\dots,\lambda_n) \neq 0,$$

whence $v(x_1,\dots,x_n) \notin V_i$ and so $v(x_1,\dots,x_n) \in V$. Contradiction. ■

7.4. THEOREM (Krop [30]). Let K be an integral domain. Then

$$\text{var}\ t_n = \begin{cases} \mathfrak{S}^n \times \mathfrak{A} & \text{if } |K^*| = \infty, \\ \mathfrak{S}^n \times \mathfrak{A}_m & \text{if } |K^*| = m. \end{cases}$$

P r o o f [71 , 78] . It is easy to see that

$$t_n = \underbrace{(K, K^*) \triangledown (K, K^*) \triangledown \dots \triangledown (K, K^*)}_{n \text{ times}} \tag{2}$$

Since (K, K^*) is a projective representation and, by 7.3, var (K, K^*) is a projective variety, we obtain from Theorem 6.2 and (2) that $\text{var}\ t_n = (\text{var}\ (K, K^*))^n$. In view of Lemma 7.3 and (1), this implies that

$$\text{var}\ t_n = \begin{cases} (\mathfrak{S} \times \mathfrak{A})^n = \mathfrak{S}^n \times \mathfrak{A} & \text{if } |K^*| = \infty, \\ (\mathfrak{S} \times \mathfrak{A}_m)^n = \mathfrak{S}^n \times \mathfrak{A}_m & \text{if } |K^*| = m. \end{cases}$$ ■

7.5. COROLLARY. Let K be an integral domain. Then the following identity forms a basis for the identities of the representation t_n:

$$(1 - [x_1, y_1])(1 - [x_2, y_2]) \dots (1 - [x_n, y_n]) \quad \text{if } |K^*| = \infty,$$

$$(1 - [x_1, y_1]z_1^m)(1 - [x_2, y_2]z_2^m) \dots (1 - [x_n, y_n]z_n^m) \quad \text{if } |K^*| = m.$$ ■

Thus the problem of describing $\mathrm{var}\ ut_n$ and $\mathrm{var}\ t_n$ is solved. We will show now that the above technique has also applications to varieties of <u>groups</u>. Namely, a description of the group varieties $\mathrm{var}\ UT_n$ and $\mathrm{var}\ T_n$ is now close at hand. To show this, recall first some standard notation. The classes of all nilpotent and of n-nilpotent groups are denoted by $\mathfrak{N}$ and $\mathfrak{N}_n$ respectively. If Θ is a class of groups, then by Θ_o we denote the class of torsion-free Θ-groups. In particular, we will use the classes $\mathfrak{N}_o$, $\mathfrak{N}_{n,o}$, $\mathfrak{A}_o$, etc. For a variety $\mathfrak{X}$ of group representations denote by $\overrightarrow{\mathfrak{X}}$ the class of all groups admitting a faithful representation in $\mathfrak{X}$. Evidently $\overrightarrow{\mathfrak{X}}$ is S- and C-closed (moreover, $\overrightarrow{\mathfrak{X}}$ is a quasivariety of groups [55]), so that $\mathrm{var}\ \overrightarrow{\mathfrak{X}} = Q\overrightarrow{\mathfrak{X}}$ (we use the same notation for closure operations on classes of groups as for the corresponding operations on classes of representations).

7.6. LEMMA. If (A, G) is a faithful representation and $\mathrm{var}\ (A, G) = \mathfrak{X}$, then $\mathrm{var}\ G = Q\overrightarrow{\mathfrak{X}}$.

P r o o f. Since $\mathfrak{X} = VQSC\{(A, G)\}$, all faithful representations from $\mathfrak{X}$ lie in the class $QSC\{(A, G)\}$, whence $\overrightarrow{\mathfrak{X}} \subseteq QSC\{G\} = \mathrm{var}\ G$ and $Q\overrightarrow{\mathfrak{X}} \subseteq \mathrm{var}\ G$. Conversely, $G \in \overrightarrow{\mathfrak{X}}$ implies that $\mathrm{var}\ G \subseteq \mathrm{var}\ \overrightarrow{\mathfrak{X}} = Q\overrightarrow{\mathfrak{X}}$. ∎

7.7. COROLLARY (i) If K is an arbitrary ring, then $\mathrm{var}\ UT_n = Q\overrightarrow{\mathfrak{S}^n}$.

(ii) If K is an integral domain, then

$$\mathrm{var}\ T_n = \begin{cases} (Q\overrightarrow{\mathfrak{S}^n})\,\mathfrak{A} & \text{if } |K^*| = \infty, \\ (Q\overrightarrow{\mathfrak{S}^n})\,\mathfrak{A}_m & \text{if } |K^*| = m. \end{cases}$$

P r o o f. (i) Apply 7.1 and 7.6. (ii) Apply 7.4, 7.6 and the following simple equalities:

$$\overrightarrow{\mathfrak{X}\times\Theta} = \overrightarrow{\mathfrak{X}}\Theta, \quad Q(\overrightarrow{\mathfrak{X}}\Theta) = (Q\overrightarrow{\mathfrak{X}})\Theta. \ \blacksquare$$

Corollary 7.7 describes the varieties $\mathrm{var}\ UT_n$ and $\mathrm{var}\ T_n$ modulo the classes of groups $\overline{\mathfrak{S}^n}$. For some "good" rings these classes are completely identified in group theoretic terms. For example, let K be a field. It is well known that:

(i) if char $K = 0$, then $\overrightarrow{\mathfrak{G}}^n$ is the class $\mathfrak{N}_{n-1,o}$ of all $(n - 1)$-nilpotent torsion-free groups (Mal'cev [41], Jennings [22]; see also P.Hall [17]);

(ii) if char $K = p$, then $\overrightarrow{\mathfrak{G}}^n$ is the variety $\mathfrak{N}_{n-1,p}$ of all groups in which the n-th term of the so-called lower p-central series is trivial (Jennings [21], Lazard [37]).

From this we obtain:

7.8.COROLLARY. Let K be an integral domain

(i) If char $K = 0$, then $\operatorname{var} UT_n = \mathfrak{N}_{n-1}$ and $\operatorname{var} T_n = \mathfrak{N}_{n-1}\mathfrak{A}$.

(ii) If char $K = p$, then $\operatorname{var} UT_n = \mathfrak{N}_{n-1,p}$ and

$$\operatorname{var} T_n = \begin{cases} \mathfrak{N}_{n-1,p}\mathfrak{A} & \text{if } |K^*| = \infty, \\ \mathfrak{N}_{n-1,p}\mathfrak{A}_m & \text{if } |K^*| = m. \end{cases}$$

P r o o f. (i) Let P be the field of fractions of K. By the preceding remark $\overrightarrow{\mathfrak{G}}^n_P = \mathfrak{N}_{n-1,o}$, whence $Q\overrightarrow{\mathfrak{G}}^n_P = \mathfrak{N}_{n-1}$ – the variety of all $(n - 1)$-nilpotent groups. Therefore $Q\overrightarrow{\mathfrak{G}}^n_K \supseteq \mathfrak{N}_{n-1}$. On the other hand, according to a well known theorem of Kaloujnine [27], $\overrightarrow{\mathfrak{G}}^n_K \subseteq \mathfrak{N}_{n-1}$ for every K. Thus $Q\overrightarrow{\mathfrak{G}}^n_K = \mathfrak{N}_{n-1}$, and it remains to apply 7.7. The proof of (ii) is analogous. ■

Of course, there are direct proofs of these formulas - for instance, see Romanovskiǐ's paper [61] in which the identities of UT_n and T_n over certain rings were first described. But in this section we wanted to show that all the main results of [13], [30] and [61] can be naturally deduced from the formula

$$\operatorname{var}(\rho_1 \nabla \rho_2) = \operatorname{var}\rho_1 \cdot \operatorname{var}\rho_2. \tag{3}$$

In conclusion let us emphasize an advantage of the ∇-approach. We will see in Appendix that the construction of triangular product can be transferred from REP-K to certain related categories. In particular, the equality (3) can be established, without any essential change in the proof, for triangular products of representations of associative and Lie algebras. This will make it possible to describe the identities of triangular matrix algebras (associative and Lie) and their canonical representations, answering questions posed in the literature (e.g., a question of Yu.N.Mal'cev in [8]).

8. Augmentation powers and dimension subgroups

Let G be a group, K a ring, and $\Delta_G = \Delta$ the augmentation ideal of the group ring KG. The powers of Δ with arbitrary transfinite exponents are defined as follows: $\Delta^1 = \Delta$, $\Delta^{\alpha+1} = \Delta^{\alpha}\cdot\Delta$, and $\Delta^{\lambda} = \bigcap_{\alpha<\lambda}\Delta^{\alpha}$ for a limit ordinal λ. There arises the augmentation series

$$KG \supseteq \Delta \supseteq \Delta^2 \supseteq \dots \Delta^{\alpha} \supseteq \Delta^{\alpha+1} \supseteq \dots \quad (1)$$

of the group ring KG. A number of well known problems is connected with this series, and an extensive bibliography is devoted to the subject. In particular, the recent book of Passi [47] deals with the powers Δ^{α} where α does not exceed the first infinite ordinal ω.

The purpose of this section is to show that certain questions connected with the augmentation series can be naturally solved by means of triangular products.

Our first question is connected with the notion of terminal. The basic ring here is the ring of integers $\mathbb{Z}$. Let G be an arbitrary group. Evidently the augmentation series in $\mathbb{Z}G$ terminates, that is, there exists an ordinal α such that $\Delta^{\alpha} = \Delta^{\alpha+1} = \dots$; the least α with this property is called the (augmentation) terminal of G and is denoted by $\tau(G)$. How to calculate $\tau(G)$ for a given group G ? It is clear that this question is too general to be completely solved.

Here we restrict ourselves to finite groups. It is well known that if G is a finite nilpotent group, then $\tau(G) = \omega$ (see [15 , 64]). Some time ago one even believed this equality to be true for every finite group. However, the conjecture turns out to be false: for each integer $n \geq 0$ there exists a finite group G such that $\tau(G) \geq \omega + n$. This was independently proved by Gruenberg and Roseblade [16] and by Kaljulaid [24] (note that in the first paper the more precise result was obtained: $\forall n\, \exists G$: $\tau(G) = \omega + n$) We will present the proof of [24] which is based on triangular products.

8.1. THEOREM. For each integer $n \geqslant 0$ there exists a finite group G such that $\tau(G) \geqslant \omega + n$.

P r o o f. Recall that for a given representation (A, G) the lower stable series

$$A = A_0 \supseteq A_1 \supseteq \dots A_\alpha \supseteq A_{\alpha+1} \supseteq \dots$$

is defined as follows: $A_{\alpha+1} = A_\alpha \circ \Delta_G = \{a \circ u \mid a \in A_\alpha,\ u \in \Delta_G$ and $A_\lambda = \bigcap_{\alpha<\lambda} A_\alpha$ for a limit ordinal λ. Clearly $A_\alpha \supseteq A \circ \Delta_G^\alpha$, but, in general, $A_\alpha \neq A \circ \Delta_G^\alpha$. In particular, the lower stable series of the regular representation $\mathrm{Reg}_{\mathbb{Z}} G = (\mathbb{Z}G, G)$ coincides with the augmentation series $\mathbb{Z}G \supseteq \Delta_G \supseteq \Delta_G^2 \supseteq \dots$. Evidently, if (B, H) is a subrepresentation of (A, G), then $B_\alpha \subseteq A_\alpha$ for every α. Further, given subsets $X \subseteq A$ and $Y \subseteq G$, the symbol $[X, Y]$ denotes the subgroup of A generated by all $x \circ (y - 1)$ with $x \in X$, $y \in Y$. Using this notation we may define the lower stable series of (A, G) as

$$A_1 = [A, G],\ A_2 = [A_1, G],\ \dots .$$

Let n be an arbitrary positive integer. Take two distinct primes p and q and consider two representations. The first is the unitriangular representation $ut_n(\mathbb{Z}_p) = (\mathbb{Z}_p^n, UT_n(\mathbb{Z}_p))$ over $\mathbb{Z}_p = \mathbb{Z}/p\mathbb{Z}$, considered as a $\mathbb{Z}$-representation, but the second is the regular representation $\mathrm{Reg}_{\mathbb{Z}} Q$ of some nontrivial finite q-group Q. Set

$$(V, G) = ut_n(\mathbb{Z}_p) \triangledown \mathrm{Reg}_{\mathbb{Z}} Q.$$

Our aim is to prove that $\tau(G) \geqslant \omega + n$.

Denote for brevity $\mathbb{Z}_p^n = A$, $UT_n(\mathbb{Z}_p) = P$, then $ut_n(\mathbb{Z}_p) = (A, P)$. Since $\mathrm{Reg}_{\mathbb{Z}} Q = (\mathbb{Z}Q, Q)$, we have

$$(V, G) = (A \oplus \mathbb{Z}Q,\ \Phi \leftthreetimes (P \times Q))$$

where $\Phi = \mathrm{Hom}_{\mathbb{Z}}(\mathbb{Z}Q, A)$. It is easy to see that Φ is isomorphic to the direct power $A^{(Q)}$. Therefore Φ is a finite group, and so is G.

Consider the lower stable series in (V, G):

$$V = V_0 \supseteq V_1 \supseteq V_2 \supseteq \dots .$$

Since $(\mathbb{Z}Q, Q) \subseteq (V, G)$, it follows that $\Delta_Q^s \subseteq V_s$ for every integer s.

(i) We show that $A \oplus \Delta_Q^s \subseteq V_s$ for every s. First let $s = 1$ and $a \in A$. Take a basis in the free abelian group $\mathbb{Z}Q$, and let b be an element of this basis. Take $\varphi \in \Phi$ such that $b\varphi = a$. Then

$$V_1 = V \circ \Delta_G \ni b\circ(\varphi - 1) = b\circ\varphi - b = b\varphi + b - b = a,$$

whence $A \subseteq V_1$.

Now let $s \geq 2$. In view of the natural decomposition $\mathbb{Z}Q = \mathbb{Z} \oplus \Delta_Q$ we have

$$\Phi = \mathrm{Hom}\,(\mathbb{Z}Q, A) = \mathrm{Hom}\,(\mathbb{Z}, A) \oplus \mathrm{Hom}\,(\Delta_Q, A), \tag{2}$$

so that $\mathrm{Hom}\,(\Delta_Q, A)$ may be regarded as a subgroup of Φ. Next, since Q is a finite q-group, the group $\Delta_Q / \Delta_Q^{s-1}$ has a nonzero exponent of the form q^m (see, e.g. [5]). Moreover, Δ_Q, being a subgroup of $\mathbb{Z}Q$, is a free abelian group. Let $x \in A$. Choose a basis element y_1 of Δ_Q and $\varphi \in \mathrm{Hom}\,(\Delta_Q, A)$ such that $y_1\varphi = x$. Since $\exp\,(\Delta_Q / \Delta_Q^{s-1}) = q^m$, it follows that $y = q^m y_1 \in \Delta_Q^{s-1}$ and

$$y\varphi = q^m (y_1\varphi) = q^m x.$$

We now prove that $A \subseteq V_s$. Take an arbitrary $a \in A$. Since A is a p-group and $p \neq q$, we can represent a in the form $a = q^m a'$, $a' \in A$. By the above, one can find $b \in \Delta_Q^{s-1}$ and $\psi \in \mathrm{Hom}\,(\Delta_Q, A)$ such that

$$a = b\psi = b\circ(\psi - 1).$$

It follows from (2) that $\psi \in G$. Therefore

$$a = b\circ(\psi - 1) \in \Delta_Q^{s-1} \circ \Delta_G \subseteq V_{s-1} \circ \Delta_G = V_s,$$

as required.

<u>(ii) $V_s = A \oplus \Delta_Q^s$ for every s, and $V_\omega = A$.</u> Indeed,

$$V_s = V \circ \Delta_G^s = A \circ \Delta_G^s + \mathbb{Z}Q \circ \Delta_G^s.$$

Since the action of G on V, regarded modulo A, coincides with that of Q, it is easy to understand that $\mathbb{Z}Q \circ \Delta_G^s \subseteq A \oplus \mathbb{Z}Q \cdot \Delta_Q^s = A \oplus \Delta_Q^s$, whence $V_s \subseteq A \oplus \Delta_Q^s$. By (i), $V_s = A \oplus \Delta_Q^s$. Since Q is a finite q-group, it follows from a theorem of Gruenberg [15] that $\bigcap_{s=1}^{\infty} \Delta_Q^s = 0$, so that $V_\omega = \bigcap_{s=1}^{\infty} V_s = A$.

<u>(iii) $A \subseteq [\mathbb{Z}Q, [\Phi, Q]]$ where $[\Phi, Q]$ means the commutator subgroup of Φ with Q in G.</u> If $b \in \mathbb{Z}Q$ and $\varphi \in \Phi$, then $b = b\circ\varphi\varphi^{-1} = (b + b\varphi)\circ\varphi^{-1} = b\circ\varphi^{-1} +$

$+ b\circ\varphi$, whence $b\circ\varphi^{-1} = b - b\varphi$. Therefore for any $g \in Q$

$$b\circ[\varphi, g] = b\circ(\varphi^{-1}g^{-1}\varphi g) = (b - b\varphi)\circ(g^{-1}\varphi g) = (b\circ g^{-1} - b\varphi)\circ(\varphi g) =$$

$$= (b\circ g^{-1} + (b\circ g^{-1})\varphi - b\varphi)\circ g = b - b\varphi + (b\circ g^{-1})\varphi,$$

$$b\circ([\varphi, g] - 1) = b - b\varphi + (b\circ g^{-1})\varphi - b = (-b + b\circ g^{-1})\varphi = (b\circ(g^{-1} - 1))\varphi.$$

The latter shows that $[\mathbb{Z}Q, [\Phi, Q]] \supseteq \{u\varphi \mid u \in \Delta_Q, \varphi \in \Phi\}$. Since Δ_Q is a free abelian group, it follows from (2) that $\{u\varphi \mid u \in \Delta_Q, \varphi \in \Phi\} = A$, whence (iii) follows.

<u>(iv) $[[\Phi, Q], Q] = [\Phi, Q]$.</u> Denote $[\Phi, Q] = \Phi_1$, then

$$\Phi \supseteq \Phi_1 \supseteq [\Phi_1, Q]. \tag{3}$$

According to the definition of the triangular product, Q acts on Φ, and it is clear that the series (3) is stable under this action. Therefore the naturally arising representation $(\Phi/[\Phi_1, Q], Q)$ is 2-stable, i.e. $(\Phi/[\Phi_1, Q])\circ\Delta_Q^2 = 0$.

Take any $x \in \Phi/[\Phi_1, Q]$, $g \in Q$. There exists k such that $g^{q^k} = 1$, whence $x\circ(g^{q^k} - 1) = 0$. On the other hand,

$$x\circ(g^{q^k} - 1) = x\circ((1 + (g - 1))^{q^k} - 1) = x\circ((1 + q^k(g - 1) + \dots) - 1) =$$

$$= x\circ(q^k(g - 1) + \dots) = q^k(x\circ((g - 1) + \dots)$$

where ... denotes a sum of terms contained in Δ_Q^2. Thus $q^k(x\circ(g - 1)) = 0$. Being a factor-group of the p-group Φ, $\Phi/[\Phi_1, Q]$ is a p-group as well. Therefore $x\circ(g - 1) = 0$, so that the representation $(\Phi/[\Phi_1, Q], Q)$ is trivial. Hence $\Phi_1 = [\Phi, Q] \subseteq [\Phi_1, Q]$, while the reverse inclusion is obvious.

Denote the subgroup $\Phi_1 \rtimes Q$ of G by H, and let I be the right ideal of $\mathbb{Z}H$ generated by all $\varphi - 1$, $\varphi \in \Phi_1$.

<u>(v) $I \subseteq \Delta_H^\omega$.</u> Indeed, in view of (iv), Φ_1 is generated by elements $[\varphi, g]$ where $\varphi \in \Phi_1$, $g \in Q$. Therefore the equality

$$[\varphi, g] - 1 = \varphi^{-1}g^{-1}\varphi g - 1 = \varphi^{-1}g^{-1}((\varphi - 1)(g - 1) - (g - 1)(\varphi - 1)) \tag{4}$$

shows that

$$I \subseteq \Delta_H^2. \tag{5}$$

It follows from (4) and (5) that $I \subseteq \Delta_H^3$. Continuing in the same manner, we obtain

$I \subseteq \cap \Delta_H^s = \Delta_H^\omega$.

(vi) The final stage. By (ii), (iii) and (v),

$$A = V_\omega \supseteq V \circ \Delta_G^\omega = (A \oplus \mathbb{Z}Q) \circ \Delta_G^\omega \supseteq \mathbb{Z}Q \circ \Delta_G^\omega \supseteq \mathbb{Z}Q \circ \Delta^\omega \supseteq \mathbb{Z}Q \circ \Delta_H^\omega \supseteq$$
$$\supseteq \mathbb{Z}Q \circ I \supseteq [\mathbb{Z}Q, \Phi_1] \supseteq A,$$

whence

$$A = V \circ \Delta_G^\omega. \tag{6}$$

Since the representation $(A, P) = ut_n(\mathbb{Z}_p)$ is stable of class n (precisely!), it follows from (6) that

$$V \circ \Delta_G^{\omega+n-1} = A \circ \Delta_G^{n-1} = A \circ \Delta_P^{n-1} \neq 0,$$
$$V \circ \Delta_G^{\omega+n} = A \circ \Delta_G^{n} = A \circ \Delta_P^{n} = 0.$$

Therefore $\Delta_G^{\omega+n-1} \neq \Delta_G^{\omega+n}$, and so $\tau(G) \geq \omega+n$. ■

The second result of this section deals with the ω-th integral dimension subgroup. Recall the necessary definitions. Let G be a group and K a commutative ring with identity. Consider the augmentation series

$$KG \supseteq \Delta \supseteq \Delta^2 \supseteq \ldots \Delta^\alpha \supseteq \Delta^{\alpha+1} \supseteq \ldots$$

and for each α set

$$\delta_\alpha(G, K) = \{g \mid g \in G, g - 1\}.$$

Clearly $\delta_\alpha(G, K)$ can be also defined as follows. If we take the representation $\mathrm{Reg}_K G$ and its factor-representation $(KG/\Delta^\alpha, G)$, then

$$\mathrm{Ker}\,(KG/\Delta^\alpha, G) = \delta_\alpha(G, K).$$

In particular, it is now clear that $\delta_\alpha(G, K)$ is a normal subgroup of G for each α. It is called the α-th dimension subgroup of G over K. If the ring K is clear from the context, we also denote $\delta_\alpha(G, K) = \delta_\alpha(G)$.

If G is a group and α an ordinal, then $\gamma_\alpha(G)$ denotes the α-th term of the lower central series of G. For more than thirty years the following dimension subgroup problem has been unanswered: does the equality

$$\delta_n(G, \mathbb{Z}) = \gamma_n(G)$$

hold for every group G and every finite n ? We omit the discussion of this problem (the interested reader is referred to [47]); note only that it has been solved in the negative by Rips [60].

However, there still remains unanswered the <u>ω-th dimension subgroup problem</u>, due to Plotkin [53] : does the equality

$$\delta_\omega(G, \mathbb{Z}) = \gamma_\omega(G)$$

hold for every group G ? We will prove here a simple but rather curious statement (Proposition 8.5) related to this problem

Following [53], we first reformulate the problem in terms of stable representations. Fix an arbitrary K, take the class $\mathfrak{S} = \mathfrak{S}_K$ of unit representations over K and consider all lower powers $\mathfrak{S}^\alpha$ (for the definition see § 11). We have an ascending series of classes of representations

$$\mathfrak{S} \subseteq \mathfrak{S}^2 \subseteq \ldots \subseteq \mathfrak{S}^\alpha \subseteq \mathfrak{S}^{\alpha+1} \subseteq \ldots$$

and, furthermore, an ascending series of classes of groups

$$\overrightarrow{\mathfrak{S}} \subseteq \overrightarrow{\mathfrak{S}^2} \subseteq \ldots \subseteq \overrightarrow{\mathfrak{S}^\alpha} \subseteq \overrightarrow{\mathfrak{S}^{\alpha+1}} \subseteq \ldots$$

(recall that $\overrightarrow{\mathfrak{X}}$ denotes the class of groups admitting a faithful representation in $\mathfrak{X}$). Since $\overrightarrow{\mathfrak{S}^\alpha}$ is a prevariety of groups (i.e. S- and C-closed), in each group G there exists the $\overrightarrow{\mathfrak{S}^\alpha}$-coradical (or $\overrightarrow{\mathfrak{S}^\alpha}$- residual), that is, a normal subgroup $(\overrightarrow{\mathfrak{S}^\alpha})^*(G) = H$ of G which is the least with respect to the property $G/H \in \overrightarrow{\mathfrak{S}^\alpha}$.

8.2. PROPOSITION. $(\overrightarrow{\mathfrak{S}^\alpha})^*(G) = \delta_\alpha(G)$ for any α and G.

P r o o f. Since $\delta_\alpha(G) = \mathrm{Ker}\,(KG/\Delta^\alpha, G)$, the representation $(KG/\Delta^\alpha, G/\delta_\alpha(G))$ is faithful. Evidently $(KG/\Delta^\alpha, G/\delta_\alpha(G)) \in \mathfrak{S}^\alpha$, whence $G/\delta_\alpha(G) \in \overrightarrow{\mathfrak{S}^\alpha}$ and so $(\overrightarrow{\mathfrak{S}^\alpha})^*(G) \subseteq \delta_\alpha(G)$.

On the other hand, let $(\overrightarrow{\mathfrak{S}^\alpha})^*(G) = H$. Then $G/H \in \overrightarrow{\mathfrak{S}^\alpha}$, so that there exists a faithful representation $(A, G/H) \in \mathfrak{S}^\alpha$ or, equivalently, a representation $(A, G) \in \mathfrak{S}^\alpha$ such that $\mathrm{Ker}\,(A, G) = H$. The group algebra KG acts on A together with G. Since $(A, G) \in \mathfrak{S}^\alpha$, it is obvious that Δ^α annihilates A. Therefore $\delta_\alpha(G) = (1 + \Delta^\alpha) \cap G$

acts identically on A, i.e. $\delta_\alpha(G) \subseteq \mathrm{Ker}(A, G) = H$.

This proposition shows that the description of $\delta_\alpha(G)$ is equivalent to the identification of the group class $\overrightarrow{\mathfrak{S}}^\alpha$ (over any ring K).

Return to the case $K = \mathbb{Z}$. Note first that if θ_1 and θ_2 are prevarieties of groups, then

$$\theta_1 = \theta_2 \Leftrightarrow \forall G : \theta_1^*(G) = \theta_2^*(G)$$

— it is evident. Denote by $\mathfrak{N}_\omega$ the prevariety of residually nilpotent ("ω-nilpotent") groups. Clearly $\mathfrak{N}_\omega^*(G) = \gamma_\omega(G)$ for every G. Therefore the ω-th dimension subgroup problem is equivalent to the following: does the equality

$$\overrightarrow{\mathfrak{S}}^\omega = \mathfrak{N}_\omega$$

hold?

It is easy to show that $\overrightarrow{\mathfrak{S}}^\omega \subseteq \mathfrak{N}_\omega$. Indeed, let $G \in \overrightarrow{\mathfrak{S}}^\omega$. Then there exists a faithful representation $(A, G) \in \mathfrak{S}^\omega$, and let

$$A = A_o \supseteq A_1 \supseteq \dots \supseteq A_n \supseteq A_{n+1} \supseteq \dots A_\omega = 0$$

be the corresponding stable series. Denote $H_n = \mathrm{Ker}(A/A_n, G)$, then $(A/A_n, G/H_n)$ is a faithful representation. By Kaloujnine's theorem [27], $G/H_n \in \mathfrak{N}_{n-1}$. Moreover, it is easy to see that $\bigcap_{n=1}^{\infty} H_n = 1$. Consequently, G is residually nilpotent.

Thus it remains to find out whether the inclusion $\overrightarrow{\mathfrak{S}}^\omega \supseteq \mathfrak{N}_\omega$ is true. Since the class $\overrightarrow{\mathfrak{S}}^\omega$ is residually closed, it is enough to prove that every nilpotent group admits a faithful ω-stable representation.

8.3. PROPOSITION. If the periodic part of a nilpotent group G admits a faithful ω-stable representation, then G also admits such a representation.

P r o o f follows immediately from two facts. The first is a theorem of Kuškuleĭ [36] asserting that if the periodic part of a nilpotent group G admits a finitely stable representation, then G admits such a representation as well. The second is that any group belonging to $\overrightarrow{\mathfrak{S}}^\omega$ is a residually $\bigcup_{n<\infty} \overrightarrow{\mathfrak{S}}^n$-group - this is evident. ∎

Since a periodic nilpotent group is the direct product of p-groups, it follows from Proposition 8.3 that the ω-th dimension subgroup problem is finally reduced to the following:

8.4. PROBLEM. Does every nilpotent p-group admit a faithful ω-stable representation over $\mathbb{Z}$?

In this connection the following result may be of interest.

8.5. PROPOSITION. Every nilpotent p-group admits a faithful $(\omega + 1)$-stable representation over $\mathbb{Z}$.

P r o o f. Denote by $\overline{\mathfrak{N}}_p$ the class of all nilpotent p-groups of finite exponent. Let G be an arbitrary nilpotent p-group and C its center. It is known that $B = G/C$ is a residually nilpotent p-group of finite exponent: $B \in A\overline{\mathfrak{N}}_p$ (see Mal'cev [41], Theorem 1 and the proof of Lemma 2). Consider the unit representation (C, 1) and the regular representation $(\mathbb{Z}B, B)$. Since $B \in A\overline{\mathfrak{N}}_p$, it follows from Theorem E of Hartley [20] that $\Delta_B^{\omega} = 0$. Therefore $(\mathbb{Z}B, B) \in \mathfrak{S}^{\omega}$, whence

$$(V, T) = (C, 1) \triangledown (\mathbb{Z}B, B) \in \mathfrak{S}^{\omega+1}.$$

It is easy to understand that the group $T = \mathrm{Hom}\,(\mathbb{Z}B, C) \leftthreetimes B$ is isomorphic to the wreath product $C \mathrm{Wr} B = C^B \leftthreetimes B$. By the Kaloujnine-Krasner Theorem G is isomorphically embedded into T. Since (V, T) is a faithful representation from $\mathfrak{S}^{\omega+1}$, it follows that $G \in \overrightarrow{\mathfrak{S}^{\omega+1}}$. ∎

From this result and the preceding observations we obtain

8.6. COROLLARY. If G is a periodic group, then

$$\delta_{\omega+1}(G, \mathbb{Z}) \subseteq \gamma_{\omega}(G). \blacksquare$$

Thus to solve the ω-th dimension subgroup problem it remains to make "only one step", so that the situation looks rather challenging.

The last result of this section deals with dimension subgroups over fields. Now the basic ring K is a field. In this case the description of dimension subgroups with finite numbers (or, equivalently, the description of the classes $\overrightarrow{\mathfrak{G}}^n$ in group-theoretic terms) is well known: as noted in Section 7,

$$\overrightarrow{\mathfrak{G}}^n = \begin{cases} \mathfrak{N}_{n-1,0} & \text{if char } K = 0, \\ \mathfrak{N}_{n-1,p} & \text{if char } K = p. \end{cases} \qquad (7)$$

Further, since $\overrightarrow{\mathfrak{G}}^\omega = A(\bigcup_{n<\infty} \overrightarrow{\mathfrak{G}}^n)$, it follows that

$$\overrightarrow{\mathfrak{G}}^\omega = \begin{cases} A\mathfrak{N}_0 & \text{if char } K = 0, \\ A\overline{\mathfrak{N}}_p & \text{if char } K = p. \end{cases} \qquad (8)$$

There naturally arises the question: to describe the classes $\overrightarrow{\mathfrak{G}}^\alpha$ for <u>all</u> ordinals α. We will give here one partial result which in essence was proved by Gringlaz and Plotkin [14].

8.6. PROPOSITION. Let K be a field. Then

$$\overrightarrow{\mathfrak{G}}^{\omega+1} = \overrightarrow{\mathfrak{G}}^{\omega+2} = \ldots = \overrightarrow{\mathfrak{G}}^{\omega 2} = \begin{cases} \mathfrak{A}_0(A\mathfrak{N}_0) & \text{if char } K = 0, \\ \mathfrak{A}_p(A\overline{\mathfrak{N}}_p) & \text{if char } K = p. \end{cases}$$

P r o o f. Evidently $\overrightarrow{\mathfrak{G}}^{\omega+1} \subseteq \overrightarrow{\mathfrak{G}}^{\omega+2} \subseteq \ldots \subseteq \overrightarrow{\mathfrak{G}}^{\omega 2}$. Let $G \in \overrightarrow{\mathfrak{G}}^{\omega 2}$, then there exists a faithful representation (V, G) with G-stable series

$$V = V_0 \supseteq V_1 \supseteq \ldots V_\omega \supseteq \ldots V_{\omega 2} = 0.$$

Denote $H_1 = \mathrm{Ker}\,(V/V_\omega, G)$, $H_2 = \mathrm{Ker}\,(V_\omega, G)$, $H = H_1 \cap H_2$. Then H acts faithfully on V and stabilizes the series $V \supseteq V_\omega \supseteq 0$, whence $H \in \overrightarrow{\mathfrak{G}}^2$. The groups G/H_1 and G/H_2 act faithfully and ω-stably on V/V_ω and V_ω respectively. Therefore G/H_1, $G/H_2 \in \overrightarrow{\mathfrak{G}}^\omega$, and since $G/H \subsetneq (G/H_1) \times (G/H_2)$, we have $G/H \in \overrightarrow{\mathfrak{G}}^\omega$. Therefore $G \in \overrightarrow{\mathfrak{G}}^2 \cdot \overrightarrow{\mathfrak{G}}^\omega$, whence, in view of (7) and (8), we obtain

$$G \in \begin{cases} \mathfrak{A}_0(A\mathfrak{N}_0) & \text{if char } K = 0, \\ \mathfrak{A}_p(A\mathfrak{N}_p) & \text{if char } K = p. \end{cases}$$

Suppose now that char $K = 0$ and let $G \in \mathfrak{A}_0(A\mathfrak{N}_0)$. Then there is a normal subgroup H of G such that $H \in \mathfrak{A}_0$ and $B = G/H \in A\mathfrak{N}_0$. Being a torsion-free abelian group, H is embedded in the vector K-space

$$H_K = K \otimes_{\mathbb{Z}} H.$$

Further, since $B \in A\mathfrak{N}_o$, it follows from [20, Theorem E] that $\Delta^{\omega} = 0$ in the group algebra KB. Therefore $\mathrm{Reg}_K B = (KB, B) \in \mathfrak{S}^{\omega}$. We proceed now as in Proposition 8.5. Take the unit representation $(H_K, 1)$ and let

$$(V, T) = (H_K, 1) \triangledown (KB, B).$$

Clearly (V, T) is a faithful and $(\omega+1)$-stable representation, whence $T \in \overrightarrow{\mathfrak{S}^{\omega+1}}$. Since $T = H_K \,\mathrm{Wr}\, B$, we have $H \,\mathrm{Wr}\, B \subsetneq T$. By the Kaloujnine-Krasner Theorem $G \subsetneq H \,\mathrm{Wr}\, B$, and so $G \subsetneq T$. Consequently, $G \in \overrightarrow{\mathfrak{S}^{\omega+1}}$.

In the case char $K = p$ the proof is similar. ■

REMARK. In [74] the full description of classes $\overrightarrow{\mathfrak{S}^{\alpha}}$ over arbitrary field and for all ordinals α has been obtained. This yields a description of all dimension subgroups δ_{α} over fields. We shall not discuss these results here since they are based not on triangular products, but on another technique.

9. The semigroup of varieties of group representations

It was pointed out in Section 1 that all varieties of group representations over a given K form a semigroup $\mathfrak{M}(K)$. In the present section we investigate the semigroup $\mathfrak{M}(K)$ in two cases: (i) K is a field, and (ii) K is a Dedekind domain. The technique of triangular products plays a major role in what follows.

If K is a field, the abstract structure of the semigroup $\mathfrak{M}(K)$ is absolutely clear in view of the following:

9.1. THEOREM (Plotkin [58]). Over any field the semigroup of varieties of group representations is free.

EQUIVALENTLY: If K is a field, then the semigroup of completely invariant ideals of KF is free.

In proving Theorem 9.1 all varieties are supposed to be nontrivial (i.e. $\neq \mathfrak{E}, \mathfrak{D}$

We begin with several preliminary results.

9.2. LEMMA. Every variety decomposes into a product of a finite number of indecomposable ones.

P r o o f. Let $\mathfrak{X}$ be a variety and $I = \operatorname{Id} \mathfrak{X}$. As usual, let Δ be the augmentation ideal of KF. It is well known that $\bigcap_{n=1} \Delta^n = 0$ [40]. Therefore there exists an integer k such that $I \subseteq \Delta^k$ but $I \nsubseteq \Delta^{k+1}$ (recall that $I \subseteq \Delta$ by Proposition 1.4).

Suppose that $\mathfrak{X}$ does not decompose into a product of a finite number of indecomposable varieties. Then $\mathfrak{X}$ can be represented as

$$\mathfrak{X} = \mathfrak{X}_1 \mathfrak{X}_2 \dots \mathfrak{X}_{k+1}$$

where $\mathfrak{X}_i$ are nontrivial varieties. If $I_i = \operatorname{Id} \mathfrak{X}_i$, then

$$I = I_{k+1} \dots I_2 I_1 \subseteq \Delta^{k+1}$$

which contradicts the above. ■

It follows that

$$\mathfrak{X}\mathfrak{Y} \neq \mathfrak{X} \neq \mathfrak{Y}\mathfrak{X} \qquad (1)$$

for any varieties $\mathfrak{X}$ and $\mathfrak{Y}$. For if, for example, $\mathfrak{X} = \mathfrak{Y}\mathfrak{X}$, then $\mathfrak{X} = \mathfrak{Y}^2\mathfrak{X} = \mathfrak{Y}^3\mathfrak{X} = \dots$ which is impossible by Lemma 9.2.

9.3. LEMMA. In the semigroup of varieties the right and left cancellation laws hold.

P r o o f. a) Suppose $\mathfrak{X}_1\mathfrak{Y} = \mathfrak{X}_2\mathfrak{Y}$ where $\mathfrak{X}_1$, $\mathfrak{X}_2$ and $\mathfrak{Y}$ are varieties. We shall prove that $\mathfrak{X}_1 = \mathfrak{X}_2$. Let $(A, H) \in \mathfrak{X}_1$ and $(E, F) = \operatorname{Fr} \mathfrak{Y}$, then

$$(V, G) = (A, H) \triangledown (E, F) \in \mathfrak{X}_1\mathfrak{Y} = \mathfrak{X}_2\mathfrak{Y},$$

whence $(\mathfrak{Y}^*(V, G), G) \in \mathfrak{X}_2$. Denote $\mathfrak{Y}^*(V, G) = W$, then it is evident that $W \subseteq A$. By Proposition 2.6,

$$(V/W, G) = (A/W, H) \triangledown (E, F),$$

so that $(A/W, H) \triangledown (E, F) \in \mathfrak{Y}$. By Theorem 6.1, var $((A/W, H) \triangledown (E, F))$ = var (A/W,

H)$\cdot \mathfrak{Y}$, whence $\mathfrak{Y} = \mathrm{var}\,(A/W, H)\cdot \mathfrak{Y}$. In view of (1), this is possible only when $(A/W, H)$ is a zero representation, i.e. $A = W$. Therefore $(A, G) = (\mathfrak{Y}_2^*(V, G), G) \in \mathfrak{X}_2$ and so $(A, H) \in \mathfrak{X}_2$. Thus $\mathfrak{X}_1 \subseteq \mathfrak{X}_2$. The reverse inclusion is proved analogously.

b) Now let $\mathfrak{Y}\mathfrak{X}_1 = \mathfrak{Y}\mathfrak{X}_2$ and assume, for example, that $\mathfrak{X}_1 \not\subseteq \mathfrak{X}_2$. Then there exists a representation $(B, H) \in \mathfrak{X}_1 \setminus \mathfrak{X}_2$. If $(E, F) = \mathrm{Fr}\,\mathfrak{Y}$, then

$$(V, G) = (E, F) \triangledown (B, H) \in \mathfrak{Y}\mathfrak{X}_1 = \mathfrak{Y}\mathfrak{X}_2 . \tag{2}$$

Since $(B, H) \in \mathfrak{X}_2$, we have $\mathfrak{X}_2^*(B, H) = B_o \neq 0$. By 1.6(ii), $\mathfrak{X}_2^*(V, G) \supseteq B_o$. Using 2.10, it is easy to see that $\mathfrak{X}_2^*(V, G) = E \oplus B_o$. Therefore (2) implies that $(E \oplus B_o, G) \in \mathfrak{Y}$. By 2.11, there exists a right epimorphism

$$(E \oplus B_o, G) \longrightarrow (E, F) \triangledown (B_o, H);$$

hence $(E, F) \triangledown (B_o, H)$ is also contained in $\mathfrak{Y}$. By 6.2, the latter representation generates the variety $\mathfrak{Y}\cdot \mathrm{var}\,(B_o, H)$. So $\mathfrak{Y} = \mathfrak{Y}\cdot \mathrm{var}\,(B_o, H)$, but this is impossible since $B_o \neq 0$. ∎

9.4. LEMMA. Let $\mathfrak{X}_1\mathfrak{X}_2 = \mathfrak{Y}_1\mathfrak{Y}_2$ where $\mathfrak{X}_i$ and $\mathfrak{Y}_i$ are varieties. If $\mathfrak{X}_2 \not\subseteq \mathfrak{Y}_2$, then there exists a nontrivial variety $\mathfrak{Z}$ such that $\mathfrak{Z}\mathfrak{Y}_2 = \mathfrak{X}_2$.

P r o o f. Let $(E_i, F) = \mathrm{Fr}\,\mathfrak{X}_i$, $i = 1,2$. Since $\mathfrak{X}_2 \not\subseteq \mathfrak{Y}_2$, we have $\mathfrak{Y}_2^*(E_2, F) = B_o \neq 0$. Denote $\mathrm{var}\,(B_o, F) = \mathfrak{Z}$, then $(E_2, F) \in \mathfrak{Z}\mathfrak{Y}_2$ and so $\mathfrak{X}_2 \subseteq \mathfrak{Z}\mathfrak{Y}_2$. On the other hand, repeating literally arguments from the previous proof, we obtain that the triangular product $(E_1, F) \triangledown (B_o, F)$ is contained in $\mathfrak{Y}_1$. Therefore

$$\mathrm{var}\,((E_1, F) \triangledown (B_o, F) = \mathfrak{X}_1\mathfrak{Z} \subseteq \mathfrak{Y}_1,$$

so that

$$\mathfrak{X}_1 \mathfrak{Z} \mathfrak{Y}_2 \subseteq \mathfrak{Y}_1 \mathfrak{Y}_2 = \mathfrak{X}_1\mathfrak{X}_2 . \tag{3}$$

Since $\mathfrak{X}_2 \subseteq \mathfrak{Z}\mathfrak{Y}_2$, we have $\mathfrak{X}_1\mathfrak{X}_2 \subseteq \mathfrak{X}_1 \mathfrak{Z} \mathfrak{Y}_2$. This and (3) imply that

$$\mathfrak{X}_1\mathfrak{X}_2 = \mathfrak{X}_1 \mathfrak{Z} \mathfrak{Y}_2 = \mathfrak{X}_1\mathfrak{X}_2 .$$

Consequently, by Lemma 9.3, $\mathfrak{X}_2 = \mathfrak{Z}\mathfrak{Y}_2$. ∎

P r o o f of Theorem 9.1 is now straightforward. By 9.2, each variety factors into a product of a finite number of indecomposable ones. Hence it suffices to show that if

$$\mathfrak{X}_1 \mathfrak{X}_2 \dots \mathfrak{X}_m = \mathfrak{Y}_1 \mathfrak{Y}_2 \dots \mathfrak{Y}_n$$

where each $\mathfrak{X}_i$ and $\mathfrak{Y}_j$ is an indecomposable variety, then $m = n$ and $\mathfrak{X}_i = \mathfrak{Y}_i$, $i = 1,2,\dots,n$.

Put $\overline{\mathfrak{X}} = \mathfrak{X}_1 \dots \mathfrak{X}_{m-1}$, $\overline{\mathfrak{Y}} = \mathfrak{Y}_1 \dots \mathfrak{Y}_{n-1}$, so that $\overline{\mathfrak{X}}\mathfrak{X}_m = \overline{\mathfrak{Y}}\mathfrak{Y}_n$. Assume that $\mathfrak{X}_m \not\subseteq \mathfrak{Y}_n$. By the previous lemma, there exists a nontrivial variety $\mathfrak{Z}$ such that $\mathfrak{Z}\mathfrak{Y}_n = \mathfrak{X}_m$. But this contradicts the indecomposability of $\mathfrak{X}_m$. Thence $\mathfrak{X}_m \subseteq \mathfrak{Y}_n$. The reverse inclusion is proved analogously, and therefore $\mathfrak{X}_m = \mathfrak{Y}_n$. By 9.3, $\overline{\mathfrak{X}} = \overline{\mathfrak{Y}}$, and the proof is completed by induction. ■

Theorem 9.1 is analogous to that of the Neumanns [44] and Šmel'kin [65] asserting that the semigroup of varieties of groups is free. Remark that the latter theorem has been reproved by Dunwoody [9] in the language of verbal subgroups which correspond one-to-one to varieties of groups. It turns out that Theorem 9.1 can be reproved in a similar way as well. Indeed, according to Cohn [7] , the group algebra KF of a free group is a fir (free ideal ring). But a theorem of Bergman and Lewin [3] states that the semigroup of ideals of a fir is free. In particular, the semigroup of all ideals of KF is free. Though, in general, a subsemigroup of a free semigroup need not be free, it has been proved in [3] that the semigroup of completely invariant ideals of KF is also free. This is equivalent to Theorem 9.1.

If the basic ring K is not a field, the structure of the semigroup $\mathfrak{M}(K)$ in general is still unknown. In any case this semigroup is not free. Indeed, let $\mathfrak{a}$ be an ideal of K and let $I_{\mathfrak{a}}$ be the set of all elements $u = \sum \lambda_i f_i$ from KF with $\lambda_i \in \mathfrak{a}$. Then $I_{\mathfrak{a}}$ is a completely invariant ideal of KF, and it is obvious that

$$I_{\mathfrak{a}} I_{\mathfrak{b}} = I_{\mathfrak{b}} I_{\mathfrak{a}} = I_{\mathfrak{a}\mathfrak{b}}$$

for any ideals $\mathfrak{a}$ and $\mathfrak{b}$ of K. This shows that $\mathfrak{M}(K)$ is not a free semigroup.(By the way, the above argument shows that the semigroup of ideals of K is anti-isomorphically embedded into $\mathfrak{M}(K)$).

Thus the problem of describing the structure of $\mathfrak{M}(K)$ for an arbitrary K is very complicated. One may really expect to solve it only if the ring K is more or less simple.

Before continuing the main topic of this section let us establish certain connections, discovered by Plotkin [56], between varieties of group representations over some ring and over its subring. Let L be a commutative ring with 1 and K its subring (with 1). Consider two maps

$$\nu : \mathfrak{M}(K) \longrightarrow \mathfrak{M}(L) \quad \text{and} \quad \nu' : \mathfrak{M}(L) \longrightarrow \mathfrak{M}(K)$$

defined as follows. If $\mathfrak{X}$ is a variety of representations over K, then $\mathfrak{X}^{\nu}$ is the class of all representations over L which as K-representations lie in $\mathfrak{X}$. Clearly $\mathfrak{X}^{\nu}$ is a variety of representations over L.

Now let $\mathfrak{Y}$ be a variety over L. Consider $\mathfrak{Y}$ as a class of representations over K and set $\mathfrak{Y}^{\nu'} = \mathrm{var}_K \mathfrak{Y}$. Evidently $\mathfrak{X}^{\nu\nu'} \subseteq \mathfrak{X}$ and $\mathfrak{Y}^{\nu'\nu} \supseteq \mathfrak{Y}$ for each $\mathfrak{X} \in \mathfrak{M}(K)$ and $\mathfrak{Y} \in \mathfrak{M}(L)$.

It is easy to find out how the ideals of identities of $\mathfrak{X}^{\nu}$ and $\mathfrak{Y}^{\nu'}$ depend on those of $\mathfrak{X}$ and $\mathfrak{Y}$. We suppose that KF is naturally contained in LF. If $\mathfrak{X} \in \mathfrak{M}(K)$ and $I = \mathrm{Id}\,\mathfrak{X}$, then LI is the ideal of LF generated by I, and it is clear that

$$\mathrm{Id}\,\mathfrak{X}^{\nu} = LI \; . \tag{4}$$

On the other hand, if $J = \mathrm{Id}\,\mathfrak{Y}$, $\mathfrak{Y}$ being a variety over L, then it is clear that

$$\mathrm{Id}\,\mathfrak{Y}^{\nu'} = J \cap KF. \tag{5}$$

From (4) we obtain

9.5. PROPOSITION. The map $\nu : \mathfrak{M}(K) \longrightarrow \mathfrak{M}(L)$ is a homomorphism of semigroups. ■

Now let K be an integral domain and P its field of fractions. Since $K \subsetneq P$, we can define the maps $\nu : \mathfrak{M}(K) \longrightarrow \mathfrak{M}(P)$ and $\nu' : \mathfrak{M}(P) \longrightarrow \mathfrak{M}(K)$ as above. A representation (A, G) over K is called _pure_ if A is a torsion-free K-module. A variety of representations over K is called pure if it is generated by pure representations. For instance, a projective variety is pure.

9.6. PROPOSITION. A variety $\mathfrak{X}$ is pure if and only if the representation $\mathrm{Fr}\,\mathfrak{X}$ is pure.

P r o o f. Since $\mathfrak{X}$ is generated by $\mathrm{Fr}\,\mathfrak{X}$, "$\mathrm{Fr}\,\mathfrak{X}$ is pure" implies "$\mathfrak{X}$ is pure". Conversely, let $\mathfrak{X}$ be a pure variety, that is, $\mathfrak{X} = \mathrm{var}\,\mathfrak{K}$, where $\mathfrak{K}$ is a class of pure representations. By 1.8,

$$\mathrm{Fr}\,\mathfrak{X} \in VSC\,\mathfrak{K},$$

but the class of all pure representations is V-, S-, and C-closed. ■

9.7. PROPOSITION. The maps

$$\nu : \mathfrak{M}(K) \longrightarrow \mathfrak{M}(P) \quad \text{and} \quad \nu' : \mathfrak{M}(P) \longrightarrow \mathfrak{M}(K)$$

yield a one-to-one correspondence between all varieties over P and all pure varieties over K.

P r o o f. By the definition of ν', $\mathfrak{Y}^{\nu'}$ is a pure variety for each $\mathfrak{Y} \in \mathfrak{M}(P)$. We will show now that for any $\mathfrak{X} \in \mathfrak{M}(K)$

$$\mathfrak{X}^{\nu\nu'} = \mathfrak{X} \iff \mathfrak{X} \text{ is pure.}$$

$\Longrightarrow$: Apply the previous remark.

$\Longleftarrow$: Let $\mathfrak{X}$ be a pure variety and $(E, F) = \mathrm{Fr}\,\mathfrak{X}$. The P-envelope $(P \otimes_K E, F) = (E_P, F)$ of (E, F), regarded as a K-representation, is contained in $\mathfrak{X}$, so that, regarded as a P-representation, it is contained in $\mathfrak{X}^{\nu}$. Since E is torsion-free, (E, F) is a K-subrepresentation of (E_P, F), whence $(E, F) \in \mathfrak{X}^{\nu\nu'}$. Therefore $\mathfrak{X} \subseteq \mathfrak{X}^{\nu\nu'}$, while the reverse inclusion is trivial.

Finally, let us show that $\mathfrak{Y}^{\nu'\nu} = \mathfrak{Y}$ for any $\mathfrak{Y} \in \mathfrak{M}(P)$. If J is the ideal of identities of $\mathfrak{Y}$ in PF, then

$$\mathrm{Id}\,(\mathfrak{Y}^{\nu'\nu}) = P(J \cap KF)$$

in view of (4) and (5). Choose an arbitrary $u \in PF$ and let λ be the common denominator for all the coefficients of u. Then $\lambda u \in J \cap KF$ and

$$u = \frac{1}{\lambda} \cdot \lambda u \in P(J \cap KF).$$

Therefore $J = P(J \cap KF)$ and $\mathfrak{Y} = \mathfrak{Y}^{\nu'\nu}$. ■

We return to our main topic. Throughout the remainder of this section K is a Dedekind domain. There are many pairwise equivalent definitions of a Dedekind domain, but the most convenient for us now is the following: it is an integral domain K such that any homomorphic image of an injective K-module is injective itself. Again let P be the field of fractions of K, and if $\rho = (A, G)$ is a representation over K, then $\rho_P = (A_P, G)$ is its P-envelope.

The main point which we wish to make here is

9.8. THEOREM. If ρ_1 and ρ_2 are pure representations over K, then

$$\mathrm{var}\,(\rho_{1P} \nabla \rho_{2P}) = \mathrm{var}\,\rho_1 \cdot \mathrm{var}\,\rho_2.$$

Before proving this theorem we deduce some immediate corollaries from it.

9.9. COROLLARY [77] . All pure varieties over K form a subsemigroup in $\mathfrak{M}(K)$. This subsemigroup is isomorphic to $\mathfrak{M}(P)$ and therefore is free.

P r o o f. Let $\mathfrak{X}$ and $\mathfrak{Y}$ be pure varieties over K and let $\rho_1 = \mathrm{Fr}\,\mathfrak{X}$, $\rho_2 = \mathrm{Fr}\,\mathfrak{Y}$. By Theorem 9.8,

$$\mathfrak{X}\mathfrak{Y} = \mathrm{var}\,\rho_1 \cdot \mathrm{var}\,\rho_2 = \mathrm{var}\,(\rho_{1P} \nabla \rho_{2P}).$$

Since the representation $\rho_{1P} \nabla \rho_{2P}$ is pure, $\mathfrak{X}\mathfrak{Y}$ is a pure variety, and the first assertion is proved.

Let $\mathfrak{M}_o(K)$ be the semigroup of all pure varieties over K. Propositions 9.5 and 9.7 imply that there is an isomorphism, namely the map $\nu' : \mathfrak{M}(P) \to \mathfrak{M}_o(K)$, between $\mathfrak{M}(P)$ and $\mathfrak{M}_o(K)$. ■

9.10. COROLLARY [77] . If I and J are completely invariant ideals of PF, then

$$IJ \cap KF = (I \cap KF)(J \cap KF)$$

P r o o f. Apply 9.9 and (5). ■

For each variety $\mathfrak{X}$ denote by $\mathfrak{X}_o$ its <u>pure part</u>, that is, the variety generated by all pure representations from $\mathfrak{X}$.

9.11. PROPOSITION [34]. The map $\mathfrak{X} \mapsto \mathfrak{X}_o$ is an endomorphism of the semigroup $\mathfrak{M}(K)$.

P r o o f. We have to prove that $(\mathfrak{X}\mathfrak{Y})_o = \mathfrak{X}_o \mathfrak{Y}_o$ for any varieties $\mathfrak{X}$ and $\mathfrak{Y}$. First we shall prove that $\mathfrak{X}_o \mathfrak{Y}_o \subseteq (\mathfrak{X}\mathfrak{Y})_o$. Since $\mathfrak{X}_o \mathfrak{Y}_o$ is pure by 9.9, it is enough to show that every pure representation from $\mathfrak{X}_o \mathfrak{Y}_o$ lies in $(\mathfrak{X}\mathfrak{Y})_o$. But it is obvious that such a representation belongs to $\mathfrak{X}\mathfrak{Y}$ and, being pure, it lies in $(\mathfrak{X}\mathfrak{Y})_o$.

The reverse inclusion is true over an arbitrary integral domain. Let (A, G) be a pure representation in $\mathfrak{X}\mathfrak{Y}$. Then there exists a G-submodule B of A such that $(B, G) \in \mathfrak{X}$ and $(A/B, G) \in \mathfrak{Y}$. Denote by $\mathrm{Is}(B) = \mathrm{Is}_A(B)$ the <u>isolator</u> of B in A:

$$\mathrm{Is}(B) \overset{\text{def.}}{=\!=} \{ x \mid (x \in A) \ \& \ (\exists \lambda \in K : \lambda x \in B)\}.$$

Clearly $\mathrm{Is}(B)$ is a G-submodule of A, so that we have a representation $(\mathrm{Is}(B), G)$. We will show that $(\mathrm{Is}(B), G) \in \mathfrak{X}$.

For every $\lambda \in K$ denote by $\frac{1}{\lambda} B$ the set of all $x \in A$ such that $\lambda x \in B$. Then $\frac{1}{\lambda} B$ is a G-submodule of A and the map $\bar{\lambda} : x \mapsto \lambda x$ is a G-monomorphism from $\frac{1}{\lambda} B$ to B. Therefore $(\frac{1}{\lambda} B, G) \in \mathfrak{X}$. Since $\mathrm{Is}(B)$ is generated by all $\frac{1}{\lambda} B$, $(\mathrm{Is}(B), G) \in \mathfrak{X}$. Since A is torsion-free, $(\mathrm{Is}(B), G) \in \mathfrak{X}_o$.

Further, $(A/\mathrm{Is}(B), G)$ is a pure representation and a homomorphic image of $(A/B, G)$. Therefore $(A/\mathrm{Is}(B), G) \in \mathfrak{Y}_o$, whence $(A, G) \in \mathfrak{X}_o \mathfrak{Y}_o$. ■

REMARKS (i) It is easy to see that if $\mathfrak{X}$ is a variety and $\mathrm{Id}\,\mathfrak{X} = I$, then

$$\mathrm{Id}\,\mathfrak{X}_o = \mathrm{Is}_{KF}(I).$$

(ii) In the case $K = \mathbb{Z}$, the statements 9.9 and 9.10 were earlier announced by Krop in [32]. Unfortunately, there are flaws in his arguments.

Theorem 9.8 is proved by the same technique as Theorem 6.2. Two lemmas are necessary; the first of them is, in fact, a modification of some arguments in [32].

9.12. LEMMA. Let $\mathfrak{X}$ and $\mathfrak{Y}$ be pure varieties and $(E, F) = \operatorname{Fr}\mathfrak{Y}$. Then there exist a free representation (L, F) of $\mathfrak{X}$ (not necessarily cyclic) and a left epimorphic image (D, F) of (L_P, F) such that

$$\operatorname{var}((D, F) \nabla (F, F)) = \mathfrak{X}\mathfrak{Y}.$$

Proof. Let $I = \operatorname{Id}\mathfrak{X}$, $J = \operatorname{Id}\mathfrak{Y}$. Denote the KF-module KF/JI by V and its KF-submodule J/JI by C. Evidently $(C, F) \in \mathfrak{X}$ and $(V/C, F) \cong (E, F)$.

Now it is convenient to use the language of KF-modules. Since $(C, F) \in \mathfrak{X}$, there exists a free representation (L, F) of $\mathfrak{X}$ with an epimorphism of KF-modules $\alpha : L \to C$. Denote $\operatorname{Ker}\alpha = L_o$. Since $\mathfrak{X}$ is pure, L is contained in its P-envelope L_P, and so $L_o \subseteq L_P$. Denote the KF-module L_P/L_o by D. Then $C = L/L_o \subseteq L_P/L_o = D$, so there is a monomorphism $\mu : C \to D$.

Being a vector space over P, L_P is an injective K-module, whence its epimorphic image D is an injective K-module as well (recall that K is a Dedekind domain). We have the diagram of KF-modules

$$\begin{array}{ccccccccc} O & \to & C & \to & V & \to & E & \to & O \\ & & \mu\downarrow & & & & & & \\ & & D & & & & & & \end{array}$$

where the row is exact. It is well known that one can construct a commutative diagram of KF-modules

$$\begin{array}{ccccccccc} O & \to & C & \to & V & \to & E & \to & O \\ & & \mu\downarrow & & \nu\downarrow & & \| & & \\ O & \to & D & \text{—} & W & \text{—} & E & \text{—} & O \end{array} \qquad (6)$$

where the second row is exact as well and ν is a monomorphism (see, e.g., S. Maclane, "Homology", Chapter III, § 1).

Now we return to the language of representations. By (6),

$$(V, F) \subseteq (W, F) \text{ and } (W/D, F) = (E, F).$$

Since D is an injective K-module, it is a direct summand of W. Therefore Theorem

3.1 can be applied and we obtain that the faithful image of (W, F) is embedded in the faithful image of $(D, F) \nabla (E, F)$. Since

$$\mathrm{Fr}\,(\mathfrak{X}\mathfrak{Y}) = (V, F) \subsetneq (W, F)$$

it follows that $\mathfrak{X}\mathfrak{Y} \subseteq \mathrm{var}\,((D, F) \nabla (E, F))$. On the other hand, (D, F) is an epimorphic image of $(L_p, F) \in \mathfrak{X}$, so $(D, F) \in \mathfrak{X}$ and $(D, F) \nabla (E, F) \in \mathfrak{X}\mathfrak{Y}$. Consequently, $\mathfrak{X}\mathfrak{Y} = \mathrm{var}\,((D, F) \nabla (E, F))$. ∎

9.13. LEMMA. Let $\mathfrak{X}$ and $\mathfrak{Y}$ be pure varieties, and let $(L, F) = \mathrm{Fr}\,\mathfrak{X}$, $(E, F) = \mathrm{Fr}\,\mathfrak{Y}$. Then

$$\mathrm{var}\,((L_p, F) \nabla (E_p, F)) = \mathfrak{X}\mathfrak{Y}.$$

P r o o f. Note first that it suffices to prove the lemma for some (not necessarily cyclic) free representation (L, F) of $\mathfrak{X}$ of countable right rank. Indeed, every such representation is obtained from the cyclic one $\mathrm{Fr}\,\mathfrak{X}$ by means of closure operations D and S, but these operations can be "carried out of triangular products" (Lemma 3.4). Thus from now on we assume (L, F) to be the free representation whose existence is assured by Lemma 9.12.

Consequently, there exists a left epimorphism $\pi : (L_p, F) \longrightarrow (D, F)$ such that

$$\mathrm{var}\,((D, F) \nabla (E, F)) = \mathfrak{X}\mathfrak{Y}. \tag{7}$$

Let us compare two triangular products $\rho_1 = (D, F) \nabla (E, F)$ and $\rho_2 = (D, F) \nabla (E_p, F)$. We have

$$\rho_1 = (D \oplus E,\ \mathrm{Hom}\,(E, D) \rightthreetimes (F \times F)),$$

$$\rho_2 = (D \oplus E_p,\ \mathrm{Hom}\,(E_p, D) \rightthreetimes (F \times F)).$$

Since $\mathfrak{Y}$ is a pure variety, E is torsion-free and so there is the natural embedding $\alpha : E \longrightarrow E_p$. Furthermore, since K is a Dedekind domain, the K-module D, being an epimorphic image of injective K-module L_p, is injective itself. Therefore the monomorphism α naturally induces the epimorphism $\alpha^* : \mathrm{Hom}\,(E_p, D) \longrightarrow \mathrm{Hom}\,(E, D)$.

It follows that there is a subrepresentation $\sigma = (D \oplus E,\ \mathrm{Hom}\,(E_p, D) \rightthreetimes (F \times F))$

of ϱ_2 and, furthermore, there is a right epimorphism of σ onto ϱ_1. Therefore var $\varrho_1 \subseteq$ var ϱ_2. Having in mind (7) and obvious inclusions $(D, F) \in \mathfrak{X}$ and $(E_P, F) \in \mathfrak{Y}$, we obtain

$$\text{var } \varrho_2 = \mathfrak{X}\mathfrak{Y}.$$

Denote $(L_P, F) \triangledown (E_P, F) = \varrho_3$. Since $\varrho_3 \in \mathfrak{X}\mathfrak{Y}$, it is enough to prove that the following inclusion holds:

$$\varrho_2 \in \text{var } \varrho_3.$$

In other words, it is enough to prove that if $u(x_1,\ldots,x_n) = \sum \lambda_k f_k(x_1,\ldots,x_n)$ is not an identity of ϱ_2, it is not an identity of ϱ_3 either.

Thus, let us suppose that $y \circ u(x_1,\ldots,x_n) = 0$ fails in $\varrho_2 = (D, F) \triangledown (E_P, F)$. Of course, we may assume that $u(x_1,\ldots,x_n)$ is an identity of (L_P, F) and (E_P, F). Then it is an identity of (D, F) and, according to (*) in § 6, there is $b \in E_P$ and $g_i = \varphi_i \bar{g}_i \bar{\bar{g}}_i$ (where $\varphi_i \in \text{Hom}(E_P, D)$, $\bar{g}_i \in F$, $\bar{\bar{g}}_i \in F$) such that

$$0 \neq b \circ u(g_1,\ldots,g_n) = \sum \lambda_k n_{ikt}(((b \circ \bar{\bar{s}}_{ikt}^{-1})\varphi_i) \circ \bar{s}_{ikt}) \circ f_k(\bar{g}_1,\ldots,\bar{g}_n) \tag{8}$$

where n_{ikt}, $\bar{\bar{s}}_{ikt}$ and $\bar{s}_{ikt}$ are defined as usual. Our aim now is to replace all φ_i by certain $\psi_i \in \text{Hom}(E_P, L_P)$.

Consider all the elements $b \circ \bar{\bar{s}}_{ikt}^{-1}$. They lie in the vector P-space E_P, so we can consider the P-subspace of E_P generated by all $b \circ \bar{\bar{s}}_{ikt}^{-1}$. Let $e_1,\ldots,e_m$ be a basis of this subspace. Multiplying by appropriate elements of P we may assume that every $b \circ \bar{\bar{s}}_{ikt}^{-1}$ is a linear combination of e_j's <u>with integral coefficients</u>, i.e. with coefficients from K.

Pick an arbitrary φ_i and construct a diagram

$$\begin{array}{ccc} & & E_P \\ & \swarrow^{\psi_i} & \downarrow \varphi_i \\ L_P & \xrightarrow{\pi} & D \end{array} \tag{9}$$

as follows. For every $j = 1,\ldots,m$ we can find an element $a_j \in L_P$ such that $a_j\pi = e_j\varphi_i$. Define a homomorphism $\psi_i : E_P \longrightarrow L_P$ by the rule: $e_j\psi_i = a_j$, $j = 1,\ldots,m$, but on "all other" elements from E_P ψ_i acts arbitrarily. Then the diagram (9) "commutes on e_j's", that is, $e_j\psi_i\pi = e_j\varphi_i$, $j = 1,\ldots,m$. Therefore this diagram "commutes on $b\circ \bar{\bar{s}}^{-1}_{ikt}$'s as well, for every $b\circ \bar{\bar{s}}^{-1}_{ikt}$ is a linear combination of e_j's with integral coefficients, but all the maps in (9) are K-homomorphisms. Thus for every $\varphi_i \in \mathrm{Hom}(E_P, D)$ we have constructed $\psi_i \in \mathrm{Hom}(E_P, L_P)$ satisfying the condition

$$(b\circ \bar{\bar{s}}^{-1}_{ikt})\psi_i\pi = (b\circ \bar{\bar{s}}^{-1}_{ikt})\varphi_i . \tag{10}$$

Finally, the elements $h_i = \psi_i \bar{g}_i \bar{\bar{g}}_i$, $i = 1,\dots,n$, lie in the acting group of the representation $\varrho_3 = (L_P, F) \triangledown (E_P, F)$. Applying the equality (*) from § 6 to these elements and the element b, we obtain:

$$b\circ u(h_1,\dots,h_n) = \sum \lambda_k n_{ikt}(((b\circ \bar{\bar{s}}^{-1}_{ikt})\psi_i)\circ \bar{s}_{ikt})\circ f_k(\bar{g}_1,\dots,\bar{g}_n). \tag{11}$$

Since $\pi : L_P \longrightarrow D$ is a homomorphism of KF-modules, it follows from (8), (10) and (11) that

$$(b\circ u(h_1,\dots,h_n))\pi = \sum \lambda_k n_{ikt}(((b\circ \bar{\bar{s}}^{-1}_{ikt})\psi_i\pi)\circ \bar{s}_{ikt})\circ f_k(\bar{g}_1,\dots,\bar{g}_n) =$$
$$= \sum \lambda_k n_{ikt}(((b\circ \bar{\bar{s}}^{-1}_{ikt})\varphi_i)\circ \bar{s}_{ikt})\circ f_k(\bar{g}_1,\dots,\bar{g}_n) = b\circ u(g_1,\dots,g_n) \neq 0.$$

Therefore $b\circ u(h_1,\dots,h_n) \neq 0$, whence $y\circ u(x_1,\dots,x_n) = 0$ fails in ϱ_3. ■

P r o o f of Theorem 9.8. Let $\varrho_1 = (A, G_1)$ and $\varrho_2 = (B, G_2)$ be pure representations, and let $\mathrm{var}\,\varrho_1 = \mathfrak{X}$, $\mathrm{var}\,\varrho_2 = \mathfrak{Y}$. Consider two classes of representations

$$\mathfrak{K} = \mathrm{VSC}\{(A_P, G_1)\} \text{ and } \mathfrak{L} = D\{(B_P, G_2)\}.$$

First we will prove that

$$\mathrm{var}(\mathfrak{K} \triangledown \mathfrak{L}) = \mathfrak{X}\mathfrak{Y}. \tag{12}$$

If $(L, F) = \mathrm{Fr}\,\mathfrak{X}$ and $(E, F) = \mathrm{Fr}\,\mathfrak{Y}$, then

$$\mathrm{var}((L_P, F) \triangledown (E_P, F)) = \mathfrak{X}\mathfrak{Y}$$

by Lemma 9.13. Therefore, to prove (12) it is enough to show that

$$(L_P, F) \triangledown (E_P, F) \in \mathrm{var}(\mathfrak{K} \triangledown \mathfrak{L}).$$

Let a bi-identity $y\circ u(x_1,\dots,x_n) = 0$ fails in $(L_P, F) \triangledown (E_P, F)$. As usual, one may assume $u(x_1,\dots,x_n)$ to be an identity of the class $\mathfrak{K}$. Let $(V, G) = (L_P, F) \triangledown (E_P, F)$. There exist $b \in E_P$ and $g_1,\dots,g_n \in G$ such that $b\circ u(g_1,\dots,g_n) \neq 0$.

Let $u(x_1,\dots,x_n) = \sum \lambda_k f_k(x_1,\dots,x_n)$ and $g_i = \varphi_i \bar{g}_i \bar{\bar{g}}_i$, then

$$b\circ u(g_i,\dots,g_n) = \sum \lambda_k n_{ikt}(((b\circ \bar{\bar{s}}_{ikt}^{-1})\varphi_i)\circ \bar{s}_{ikt})\circ f_k(\bar{g}_1,\dots,\bar{g}_n).$$

Denote by M the P-subspace of E_P generated by all $b\circ \bar{\bar{s}}_{ikt}^{-1}$. Since B is torsion-free, it follows from Lemma 6.5 that there exists a homomorphism

$$\nu : (E_P, F) \longrightarrow (B_P, G_2)^m$$

which is one-to-one on M. Then M, being a subspace of the vector space B_P^m, is a direct summand of B_P^m. Therefore one can find a P-linear map $\eta : B_P^m \longrightarrow E_P$ such that $\nu\eta$ acts on M identically:

$$\forall x \in M : \quad x^{\nu\eta} = x. \tag{13}$$

Suppose now that $u(x_1,\dots,x_n)$ is an identity of the class $\mathfrak{K} \triangledown \mathfrak{L}$. Using Proposition 1.8, it is easy to understand that $(L\ , F) \in \mathfrak{K}$. It follows that $u(x_1,\dots,x_n)$ must be an identity of

$$(L_P, F) \triangledown (B_P, G_2)^m = (L_P \oplus B_P^m, \operatorname{Hom}(B_P^m, L_P) \leftthreetimes (F \times G_2^m)). \tag{14}$$

Denote the product of maps $\eta\varphi_i$ by ψ_i. Then $\psi_i \in \operatorname{Hom}(B_P^m, L_P)$ and, by (13),

$$(b\circ \bar{\bar{s}}_{ikt}^{-1})^{\nu}\psi_i = (b\circ \bar{\bar{s}}_{ikt}^{-1})^{\nu\eta}\varphi_i = (b\circ \bar{\bar{s}}_{ikt}^{-1})\varphi_i.$$

Furthermore, let $c = b^{\nu}$, $h_i = \bar{\bar{g}}_i^{\nu}$, $\tilde{s}_{ikt} = s_{ikt}(h_1,\dots,h_n)$, and $t_i = \psi_i \bar{g}_i h_i$; then

$$c \in B_P^m,\ h_i \in G_2^m,\ s_{ikt} \in G_2^m,\ t_i \in \operatorname{Hom}(B_P^m, L_P) \leftthreetimes (F \times G_2^m).$$

Repeating literally calculations on p. 54, we obtain that in the triangular product (14)

$$c\circ u(t_1,\dots,t_n) = \dots = b\circ u(g_1,\dots,g_n) \neq 0.$$

Thus $u(x_1,\dots,x_n)$ is not an identity of the class $\mathfrak{K} \triangledown \mathfrak{L}$, whence (12) follows.

Finally, by Lemma 3.4 and (12),

$$\mathfrak{X}\mathfrak{Y} = \operatorname{var}(\mathfrak{K} \triangledown \mathfrak{L}) = \operatorname{var}((VSC(A_P, G_1)) \triangledown (D(B_P, G_2))) \subseteq$$
$$\subseteq \operatorname{var}((A_P, G_1) \triangledown (B_P, G_2)),$$

but the reverse inclusion is trivial. ■

9.14. PROBLEM. Does Theorem 9.8 remain true over an arbitrary integral domain ?

We expect this problem to be answered in the affirmative.

In particular, the above results hold over the ring of integers. They give an essential information about the semigroup $\mathcal{M}(\mathbb{Z})$, but the general problem of describing the structure of this semigroup ([55] , problem 16) still remains unsolved. In this connection we note one more result concerning $\mathcal{M}(\mathbb{Z})$.

9.15. PROPOSITION [31] . All projective varieties over $\mathbb{Z}$ form a subsemigroup of $\mathcal{M}(\mathbb{Z})$.

P r o o f. Let $\mathfrak{X}_1$ and $\mathfrak{X}_2$ be projective varieties over $\mathbb{Z}$, and let $(E_i, F) = \operatorname{Fr} \mathfrak{X}_i$. If

$$(A, G) = (E_1, F) \triangledown (E_2, F)$$

then, by 6.3, var $(A, G) = \mathfrak{X}_1\mathfrak{X}_2$. Now let $(E, F) = \operatorname{Fr}(\mathfrak{X}_1\mathfrak{X}_2)$. It follows from 1.8 that

$$(E, F) \in \quad \operatorname{VSC}\{(A, G)\}. \tag{15}$$

Since $\mathfrak{X}_1$ and $\mathfrak{X}_2$ are projective, E_1 and E_2 are free abelian groups, and so is $A = E_1 \oplus E_2$. By (15), E is a subgroup of some Cartesian power of A. Since E is a countable group, it follows from a well known theorem of the theory of abelian groups that E is free abelian. Therefore $\mathfrak{X}_1\mathfrak{X}_2$ is a projective variety. ∎

We do not know whether this fact remains true over an arbitrary integral domain. Clearly the answer is "yes" if Problem 6.9 is solved in the affirmative.

NOTE. The equality

$$IJ \cap KF = (I \cap KF)(J \cap KF)$$

can be proved in greater generality than it has been done in Corollary 9.10. Namely, G.M.Bergman has shown that it is true if K is a Dedekind domain, I a right ideal, but J a left ideal of PF. On the other hand, he has given a short and elegant example showing that over an arbitrary domain this equality need not hold even for two-sided ideals. The author is grateful to G.M.Bergman for pointing out these results.

10. The semigroups of radical classes and prevarieties of group representations

In the remainder of Chapter 2 the basic ring K is a field though this requirement is sometimes immaterial.

The problems considered in this section arose by analogy with two well known group theoretic problems which are still unsolved.

(i) Is the semigroup of completely decomposable (hereditary) radical classes of groups free (Plotkin [48]) ?

(ii) Is the semigroup of completely decomposable prevarieties of groups free (Tsalenko [66]) ?

Similar questions can, in a natural way, be posed for representations. But in contrast to (i) and (ii), they have been solved. It is our nearest aim to present the corresponding solutions based entirely on the technique of triangular products. Unless otherwise specified, the results of this section are due to Vovsi [69, 70, 71, 75].

A class of representations is called <u>radical</u> (more precisely, hereditary radical [51]) if it is V-, Q-, S-, and C_r-closed. For any class of representations $\mathfrak{X}$ denote by rad $\mathfrak{X}$ the radical class generated by $\mathfrak{X}$. Evidently rad is a closure operation and, by 1.3,

$$\text{rad} = VQSC_r.$$

If $\mathfrak{X}$ is a radical class and (V, G) an arbitrary representation, then it is easy to see that V contains a G-submodule A such that $(A, G) \in \mathfrak{X}$ and any other G-submodule of V with this property is contained in A. This G-submodule A is called the <u>$\mathfrak{X}$-radical</u> of (V, G) and is denoted by $\mathfrak{X}(V, G)$.

A class of representation is called a <u>prevariety</u> if it is V-, $O_{\bar{r}}$, S-, and C-closed. Let pvar $\mathfrak{X}$ denote the prevariety generated by an arbitrary class of representations $\mathfrak{X}$; then pvar is also a closure operation and, by 1.3,

$$\mathrm{pvar} = VQ_r SC.$$

If $\mathfrak{X}$ is a prevariety and (V, G) a representation, it is easy to see that V contains a G-submodule A such that $(V/A, G) \in \mathfrak{X}$ and A is contained in each G-submodule of V with this property. A is called the <u>$\mathfrak{X}$-coradical</u> of (V, G) and is denoted by $\mathfrak{X}^*(V, G)$.

Since a variety of group representations can be defined as a class which is closed under the operations V, Q, S and C, it follows from the definitions:

10.1. PROPOSITION. A class of representations is a variety if and only if it is both a radical class and a prevariety. ■

We see that radical classes and prevarieties are mutually dual generalizations of varieties. In particular, if $\mathfrak{X}$ is a variety, then in every representation (V, G) there exist both the $\mathfrak{X}$-radical $\mathfrak{X}(V, G)$ and the $\mathfrak{X}$-coradical (= $\mathfrak{X}$-verbal) $\mathfrak{X}^*(V, G)$. For instance, consider the variety $\mathfrak{S}^n$ of n-stable representations, and let (V, G) be any representation. Then the $\mathfrak{S}^n$-radical of (V, G) is the n-th term of the upper stable series of (V, G), but the $\mathfrak{S}^n$-coradical is the n-th term of the lower stable series.

The following proposition is straightforward.

10.2. PROPOSITION. The product of radical classes (prevarieties) is a radical class (prevariety). ■

Therefore we may speak of semigroups of radical classes and prevarieties of group representations over a given K, keeping in mind, of course, that these "semigroups" are defined not on sets of elements, but on classes. We shall denote these semigroups by $\mathcal{R}(K) = \mathcal{R}$ and $\mathcal{P}(K) = \mathcal{P}$ respectively. Let us emphasize that, by 10.1,

$$\mathcal{R}(K) \cap \mathcal{P}(K) = \mathcal{M}(K) \qquad (1)$$

Since $\mathcal{M}(K)$ is a free semigroup (theorem 9.1), there naturally arises the question: <u>over an arbitrary field, are the semigroups of radical classes and prevarieties</u>

free ? The answer, generally speaking, turns out to be in the negative, mainly because each of these semigroups is not generated by its indecomposable elements (this will be proved a little later).

We will now consider only those radical classes (prevarieties) which decompose into a product of a finite number if indecomposable ones. Such radical classes (prevarieties) are called completely decomposable; evidently they form a subsemigroup in $\mathcal{R}$ (in $\mathcal{P}$).

Our main objective here is to prove the following two theorems (Vovsi [71]).

10.3. THEOREM. Over any field, the semigroup of completely decomposable radical classes of group representations is free.

10.4. THEOREM. Over any field, the semigroup of completely decomposable prevarieties of group representations is free.

In what follows Theorem 9.3 will be proved first. Then Theorem 9.4 will be proved, but the rest of this section will be concerned with deducing several corollaries and some relevant comments.

For brevity, from now on we shall often say "radical" instead of "radical class". To prove Theorem 10.3, several preliminary results have to be established. The first of them has been proved in [50] ; it is analogous to a well known property of radicals in wreath products of groups [67] .

10.5. LEMMA. Let $(V, G) = (A, G_1) \triangledown (B, G_2)$ and let $\mathfrak{X}$ be a radical class. If $\mathfrak{X}(A, G_1) \subset A$, then $\mathfrak{X}(V, G) = \mathfrak{X}(A, G_1)$.

P r o o f. By Corollary 2.10, either $\mathfrak{X}(V, G) \subset A$ or $A \subsetneq \mathfrak{X}(V, G)$. Suppose the second. Then the subrepresentation (A, G) of (V, G) belongs to $\mathfrak{X}$. Since there exists a right epimorphism $(A, G) \longrightarrow (A, G_1)$, we have $A = \mathfrak{X}(A, G) = \mathfrak{X}(A, G_1)$ which contradict the hypothesis. Therefore $\mathfrak{X}(V, G) \subset A$, and so

$$\mathfrak{X}(V, G) = \mathfrak{X}(A, G) = \mathfrak{X}(A, G_1).$$

10.6. LEMMA. Let $\mathfrak{X}_1 \mathfrak{X}_2 \subseteq \mathfrak{Y}_1 \mathfrak{Y}_2$, where $\mathfrak{X}_i$ and $\mathfrak{Y}_i$ are radicals. If $\mathfrak{X}_1 \nsubseteq \mathfrak{Y}_1$, then there exists a nontrivial radical $\mathfrak{Z}$ such that $\mathfrak{Z}\mathfrak{X}_2 \subseteq \mathfrak{Y}_2$.

Proof. a) We claim that $\mathfrak{X}_2 \subseteq \mathfrak{Y}_2$. For if $(A, G_1) \in \mathfrak{X}_1 \setminus \mathfrak{Y}_1$ and $(B, G_2) \in \mathfrak{X}_2$, then

$$(V, G) = (A, G_1) \triangledown (B, G_2) \in \mathfrak{X}_1 \mathfrak{X}_2 = \mathfrak{Y}_1 \mathfrak{Y}_2,$$

whence $(V/\mathfrak{Y}_1(V, G), G) \in \mathfrak{Y}_2$. Let $\mathfrak{Y}_1(A, G_1) = A_o$, then $A_o \subset A$ and Lemma 10.5 imply $\mathfrak{Y}_1(V, G) = A_o$. Therefore there exists an evident epimorphism of $(V/\mathfrak{Y}_1(V, G), G)$ onto (B, G_2), whence $(B, G_2) \in \mathfrak{Y}_2$. This shows that $\mathfrak{X}_2 \subseteq \subseteq \mathfrak{Y}_2$, as claimed.

b) Let $\mathfrak{Z}_o$ be the class of all representations $(A/\mathfrak{Y}_1(A, G), G)$, (A, G) being an arbitrary representation of $\mathfrak{X}_1$. Since $\mathfrak{X}_1 \nsubseteq \mathfrak{Y}_1$, it follows that $\mathfrak{Z}_o \neq \mathfrak{E}$, whence $\mathfrak{Z} = \text{rad}\, \mathfrak{Z}_o \neq \mathfrak{E}$. It is not hard to see that

$$V\mathfrak{Z}_o = \mathfrak{Z}_o, \quad C_r\mathfrak{Z}_o = \mathfrak{Z}_o. \tag{2}$$

Indeed, let, for example, $(W, G) \in \mathfrak{Z}_o$ and let $\mu : (W, H) \longrightarrow (W, G)$ be a right epimorphism. By the definition of $\mathfrak{Z}_o$, there exists $(A, G) \in \mathfrak{X}_1$ such that $(W, G) = (A/\mathfrak{Y}_1(A, G), G)$. The group homomorphism $\mu : H \longrightarrow G$ enables us to define a representation (A, H) and a right epimorphism $\bar{\mu} : (A, H) \longrightarrow (A, G)$. Since a radical class is V-closed,

$$\mathfrak{Y}_1(A, H) = \mathfrak{Y}_1(A, G)$$

so that $(W, H) = (A/\mathfrak{Y}_1(A, H), H)$. Since $(A, G) \in \mathfrak{X}_1$ and $V\mathfrak{X}_1 = \mathfrak{X}_1$, we have $(A, H) \in \mathfrak{X}_1$. Therefore $(W, H) \in \mathfrak{Z}_o$, whence $V\mathfrak{Z}_o = \mathfrak{Z}_o$. The rest of (2) is proved analogously.

c) Let us prove that

$$\mathfrak{Z}_o \triangledown \mathfrak{X}_2 \subseteq \mathfrak{Y}_2. \tag{3}$$

Let $(A, G_1) \in \mathfrak{Z}_o$, $(B, G_2) \in \mathfrak{X}_2$, and let $(V, G) = (A, G_1) \triangledown (B, G_2)$. There exists $(M, G_1) \in \mathfrak{X}_1$ such that $(M/\mathfrak{Y}_1(M, G_1), G_1) = (A, G_1)$. Then

$$(\bar{V}, \bar{G}) = (M, G_1) \triangledown (B, G_2) \in \mathfrak{X}_1 \mathfrak{X}_2 = \mathfrak{Y}_1 \mathfrak{Y}_2,$$

whence $(\bar{V}/\mathcal{Y}_1(\bar{V}, \bar{G}), \bar{G}) \in \mathcal{Y}_2$.

If $A = 0$, then $(V, G) \in \mathfrak{X}_2$ and, by a), $(V, G) \in \mathcal{Y}_2$, as required. If $A \neq 0$, then $M_o = \mathcal{Y}_1(M, G_1)$ must be a proper subspace of M. By Lemma 10.5, $\mathcal{Y}_1(\bar{V}, \bar{G}) = M_o$. By 2.6 (ii), the epimorphism $(M, G_1) \to (M/M_o, G_1) = (A, G_1)$ can be extended to an epimorphism $\mu : (\bar{V}, \bar{G}) \to (V, G)$. It is clear that the subspace M_o is annihilated by μ ; hence we can "pass" μ through $(\bar{V}/M_o, \bar{G})$. Consequently, $(\bar{V}/M_o, \bar{G}) \in \mathcal{Y}_2$ implies $(V, G) \in \mathcal{Y}_2$, as required.

d) From (2), (3) and Lemma 3.4

$$\mathfrak{Z} \triangledown \mathfrak{X}_2 = (\mathrm{rad}\, \mathfrak{Z}_o) \triangledown \mathfrak{X}_2 = (VQSC_r \mathfrak{Z}_o) \triangledown \mathfrak{X}_2 =$$
$$= (VQS\mathfrak{Z}_o) \triangledown \mathfrak{X}_2 \subseteq VQS(\mathfrak{Z}_o \triangledown \mathfrak{X}_2) \subseteq \mathcal{Y}_2,$$

whence, by Lemma 3.5, we obtain

$$\mathfrak{Z}\mathfrak{X}_2 \subseteq VSQ_r(\mathfrak{Z} \triangledown \mathfrak{X}_2) \subseteq \mathcal{Y}_2. \blacksquare$$

10.7. PROPOSITION. Let $\mathfrak{X}_1\mathcal{Y} = \mathfrak{X}_2\mathcal{Y}$, where $\mathfrak{X}_i$ are radicals, $\mathcal{Y}$ is a completely decomposable radical. Then $\mathfrak{X}_1 = \mathfrak{X}_2$.

P r o o f. We may assume that $\mathcal{Y}$ is indecomposable. For if we can cancel all indecomposable right factors, we can cancel completely decomposable right factors as well.

Suppose that $\mathfrak{X}_1 \neq \mathfrak{X}_2$. For example, let $\mathfrak{X}_1 \not\subseteq \mathfrak{X}_2$. By Lemma 10.6 there exists a nontrivial radical $\mathfrak{Z}$ such that $\mathfrak{Z}\mathcal{Y} \subseteq \mathcal{Y}$. Hence $\mathfrak{Z}\mathcal{Y} = \mathcal{Y}$, but this contradicts the indecomposability of $\mathcal{Y}$. $\blacksquare$

10.8. PROPOSITION. Let $\mathcal{Y}\mathfrak{X}_1 = \mathcal{Y}\mathfrak{X}_2$, where $\mathfrak{X}_i$ are radicals and $\mathcal{Y}$ is a completely decomposable radical. Then $\mathfrak{X}_1 = \mathfrak{X}_2$.

P r o o f. As in the proof of Proposition 10.7, we may assume $\mathcal{Y}$ to be indecomposable. Suppose $\mathfrak{X}_1 \not\subseteq \mathfrak{X}_2$. Let $(A, G_1) \in \mathcal{Y}$ and $(B, G_2) \in \mathfrak{X}_1 \setminus \mathfrak{X}_2$, then

$$(V, G) = (A, G_1) \triangledown (B, G_2) \in \mathcal{Y}\mathfrak{X}_1 = \mathcal{Y}\mathfrak{X}_2,$$

whence $(V/\mathcal{Y}(V, G), G) \in \mathfrak{X}_2$. Since $(B, G_2) \notin \mathfrak{X}_2$, it follows that $\mathcal{Y}(V,$

G) $\supset$ A, i.e. $\mathfrak{Y}(V, G) = A \oplus B_o$, where $0 \neq B_o \subseteq B$. The representation $(A \oplus B_o, G)$ belongs to $\mathfrak{Y}$ and, by Lemma 2.11, there exists a right epimorphism $(A \oplus B_o, G) \to (A, G_1) \triangledown (B_o, G_2)$; hence the latter also belongs to $\mathfrak{Y}$.

Let $(K, 1)$ be the one-dimensional representation of the trivial group. Of course, it is embeddable into (B_o, G_2). By Lemma 3.5 (n), $(A, G_1) \triangledown (K, 1) \in \mathfrak{Y}$. Since (A, G_1) is an arbitrary object from $\mathfrak{Y}$,

$$\mathfrak{Y} \triangledown \{(K, 1)\} \subseteq \mathfrak{Y}. \tag{4}$$

It is evident that $\mathrm{rad}\{(K, 1)\} = \mathfrak{S}$. Therefore, using (4), Lemmas 3.4 and 3.5 we obtain

$$\begin{aligned} \mathfrak{Y} \triangledown \mathfrak{S} = \mathfrak{Y} \triangledown \mathrm{rad}\{(K, 1)\} = \mathfrak{Y} \triangledown VQSC_r\{(K, 1)\} \subseteq \\ \subseteq \mathrm{rad}\,(\mathfrak{Y} \triangledown \{(K, 1)\}) \subseteq \mathfrak{Y}. \end{aligned} \tag{5}$$

By Lemma 3.5 (k) and (5),

$$\mathfrak{Y}\mathfrak{S} \subseteq VSQ_r(\mathfrak{Y} \triangledown \mathfrak{S}) \subseteq \mathfrak{Y},$$

whence $\mathfrak{Y}\mathfrak{S} = \mathfrak{Y}$. This contradicts the indecomposability of $\mathfrak{Y}$. Therefore $\mathfrak{X}_1 \subseteq \subseteq \mathfrak{X}_2$ and analogously $\mathfrak{X}_2 \subseteq \mathfrak{X}_1$. ■

10.9. PROPOSITION. Let $\mathfrak{X}_1\mathfrak{X}_2 = \mathfrak{Y}_1\mathfrak{Y}_2$ where $\mathfrak{X}_i$ and $\mathfrak{Y}_i$ are radicals. If $\mathfrak{X}_1 \not\subseteq \mathfrak{Y}_1$, then there exists a nontrivial radical $\mathfrak{Z}$ such that

$$\mathfrak{X}_1\mathfrak{X}_2 = \mathfrak{Y}_1\mathfrak{Z}\mathfrak{X}_2 = \mathfrak{Y}_1\mathfrak{Y}_2.$$

Moreover, if $\mathfrak{X}_2$ is completely decomposable, then $\mathfrak{X}_1 = \mathfrak{Y}_1\mathfrak{Z}$, but if $\mathfrak{Y}_1$ is completely decomposable, then $\mathfrak{Z}\mathfrak{X}_2 = \mathfrak{Y}_2$.

P r o o f. As in the proof of Lemma 10.6, let

$$\mathfrak{Z}_o = \{(A/\mathfrak{Y}_1(A, G), G) \mid (A, G) \in \mathfrak{X}_1\}, \quad \mathfrak{Z} = \mathrm{rad}\,\mathfrak{Z}_o;$$

then $\mathfrak{Z}$ is a nontrivial radical and

$$\mathfrak{Z}\mathfrak{X}_2 \subseteq \mathfrak{Y}_2. \tag{6}$$

By the definition of $\mathfrak{Z}_o$, $\mathfrak{X}_1 \subseteq \mathfrak{Y}_1\mathfrak{Z}_o$ and so

$$\mathfrak{X}_1 \subseteq \mathfrak{Y}_1\mathfrak{Z}. \tag{7}$$

Since the multiplication of classes of representations is associative, it follows from

(6) and (7) that

$$\mathfrak{X}_1 \mathfrak{X}_2 \subseteq \mathfrak{Y}_1 \mathfrak{Z} \mathfrak{Y}_2 \subseteq \mathfrak{Y}_1 \mathfrak{Y}_2. \tag{8}$$

By hypothesis, $\mathfrak{X}_1 \mathfrak{X}_2 = \mathfrak{Y}_1 \mathfrak{Y}_2$. This shows that in fact there are equalities in (8). The remainder follows immediately from Propositions 10.7 and 10.8. ■

P r o o f of Theorem 10.3. Let

$$\mathfrak{X}_1 \mathfrak{X}_2 \dots \mathfrak{X}_m = \mathfrak{Y}_1 \mathfrak{Y}_2 \dots \mathfrak{Y}_n,$$

where $\mathfrak{X}_i$ and $\mathfrak{Y}_i$ are indecomposable radicals. Denote $\overline{\mathfrak{X}} = \mathfrak{X}_2 \dots \mathfrak{X}_m$, $\overline{\mathfrak{Y}} = \mathfrak{Y}_2 \dots \mathfrak{Y}_n$, then $\mathfrak{X}_1 \overline{\mathfrak{X}} = \mathfrak{Y}_1 \overline{\mathfrak{Y}}$. Assume $\mathfrak{X}_1 \not\subseteq \mathfrak{Y}_1$. By Proposition 10.9, there exists a nontrivial radical $\mathfrak{Z}$ such that $\mathfrak{X}_1 = \mathfrak{Y}_1 \mathfrak{Z}$. Since $\mathfrak{X}_1$ is indecomposable, our assumption is false and $\mathfrak{X}_1 \subseteq \mathfrak{Y}_1$. The reverse inclusion is proved analogously; therefore $\mathfrak{X}_1 = \mathfrak{Y}_1$ and, by Proposition 10.8, $\overline{\mathfrak{X}} = \overline{\mathfrak{Y}}$. The proof is completed by induction. ■

Theorem 10.4 is proved in a dual manner and we try to emphasize this duality whenever it is possible.

10.10. LEMMA. Let $(V, G) = (A, G_1) \triangledown (B, G_2)$ and let $\mathfrak{X}$ be a prevariety. If $\mathfrak{X}^*(B, G_2) \supset 0$, then $\mathfrak{X}^*(V, G) = A \oplus \mathfrak{X}^*(B, G_2)$.

P r o o f. It is easy to show that the function $\mathfrak{X}^*$ is hereditary, that is, $\rho_1 \subseteq \rho_2 \Rightarrow \mathfrak{X}^*(\rho_1) \subseteq \mathfrak{X}^*(\rho_2)$. Therefore $\mathfrak{X}^*(B, G_2) \subseteq \mathfrak{X}^*(V, G)$. It follows that $\mathfrak{X}^*(V, G) \not\subseteq A$ whence, by 2.10, $A \subseteq \mathfrak{X}^*(V, G)$. Thus

$$A \oplus \mathfrak{X}^*(B, G_2) \subseteq \mathfrak{X}^*(V, G).$$

Furthermore, $A \oplus \mathfrak{X}^*(B, G_2)$ is a G-submodule of V and

$$(V/(A \oplus \mathfrak{X}^*(B, G_2)), G) \sim (B/\mathfrak{X}^*(B, G_2), G_2) \in \mathfrak{X},$$

whence $\mathfrak{X}^*(V, G) \subseteq A \oplus \mathfrak{X}^*(B, G_2)$. ■

10.11. LEMMA. Let $\mathfrak{X}\mathfrak{Y}_1 \subseteq \mathfrak{X}\mathfrak{Y}_2$ where $\mathfrak{X}$ and $\mathfrak{Y}_2$ are prevarieties and $\mathfrak{Y}_1$ is a $\langle V, D \rangle$-closed class of representations. If $\mathfrak{Y}_1 \not\subseteq \mathfrak{Y}_2$, then $\mathfrak{X}\mathfrak{S} = \mathfrak{X}$.

P r o o f. We define a class $\mathfrak{Z}_o$ of representations as follows: $(W, G) \in \mathfrak{Z}_o$

if and only if there exists a representation $(V, G) \in \mathfrak{Y}_1$ such that $(\mathfrak{Y}_2^*(V, G), G) = (W, G)$. Since $\mathfrak{Y}_1 \not\subseteq \mathfrak{Y}_2$, it follows that $\mathfrak{Z}_o \neq \mathfrak{E}$.

a) We state that $D\mathfrak{Z}_o = \mathfrak{Z}_o$ and $V\mathfrak{Z}_o = \mathfrak{Z}_o$. Indeed, let $(W, G) \in \mathfrak{Z}_o$ and let $\mu : (W, H) \longrightarrow (W, G)$ be a right epimorphism. Pick a representation $(V, G) \in \mathfrak{Y}_1$ such that $(\mathfrak{Y}_2^*(V, G), G) = (W, G)$. The epimorphism of groups $\mu : H \longrightarrow G$ allows us to define the representation (V, H) and the right epimorphism $\mu' : (V, H) \longrightarrow (V, G)$. It is obvious that since $\mathfrak{Y}_2^*(V, G) = W$, we have also $\mathfrak{Y}_2^*(V, H) = W$. Hence $(W, H) \in \mathfrak{Z}_o$ by the definition of $\mathfrak{Z}_o$.

Now suppose $(W_i, G_i) \in \mathfrak{Z}_o$, $i \in I$. We will prove that their direct product $\prod_{i \in I}(W_i, G_i)$ is also contained in $\mathfrak{Z}_o$. For each $i \in I$ there exists a representation $(V_i, G_i) \in \mathfrak{Y}_1$ such that $(\mathfrak{Y}_2^*(V_i, G_i), G_i) = (W_i, G_i)$. Let $(V, G) = \prod(V_i, G_i)$. It is easy to see that the coradical $\mathfrak{Y}_2^*$ commutes with direct products, and hence

$$(\mathfrak{Y}_2^*(V, G), G) = \prod(W_i, G_i).$$

Since $(V, G) \in \mathfrak{Y}_1$, we have $\prod(W_i, G_i) \in \mathfrak{Z}_o$, as required.

b) Let us prove that $\mathfrak{X} \triangledown \mathfrak{Z}_o \subseteq \mathfrak{X}$. Suppose $(A, G_1) \in \mathfrak{X}$, $(B, G_2) \in \mathfrak{Z}_o$ and $(V, G) = (A, G_1) \triangledown (B, G_2)$. There exists $(C, G_2) \in \mathfrak{Y}_1$ such that $(\mathfrak{Y}_2^*(C, G_2), G_2) = (B, G_2)$. Then

$$(V', G') = (A, G_1) \triangledown (C, G_2) \in \mathfrak{X}\mathfrak{Y}_1 = \mathfrak{X}\mathfrak{Y}_2,$$

whence $(\mathfrak{Y}_2^*(V', G'), G') \in \mathfrak{X}$. Furthermore, $B = \mathfrak{Y}_2^*(C, G_2) \subseteq \mathfrak{Y}_2^*(V', G')$. By Lemma 10.10, $\mathfrak{Y}_2^*(V', G) = A \oplus B$ and we see that $(A \oplus B, G') \in \mathfrak{X}$. By Lemma 2.11, there exists a right epimorphism of $(A \oplus B, G')$ onto (V, G). Hence $(V, G) \in \mathfrak{X}$, as required.

c) It is clear that if (V, G) is a nonzero representation, then $\mathfrak{G} \in VD\{(V,G)\}$. Since $\mathfrak{Z}_o \neq \mathfrak{E}$, it follows that

$$\mathfrak{X} \triangledown \mathfrak{G} \subseteq \mathfrak{X} \triangledown (VD\mathfrak{Z}_o). \tag{9}$$

Using Lemmas 3.4 and 3.5, and also the above observations, we have

$$\mathfrak{X} \triangledown (V D \mathfrak{Z}_o) \subseteq V(\mathfrak{X} \triangledown D \mathfrak{Z}_o) = V(\mathfrak{X} \triangledown \mathfrak{Z}_o) \subseteq V \mathfrak{X} = \mathfrak{X}.$$

Because of (9), it follows that $\mathfrak{X} \triangledown \mathfrak{G} \subseteq \mathfrak{X}$. Consequently, from Lemma 3.5 (k) we have

$$\mathfrak{X} \mathfrak{G} \subseteq VSQ_r(\mathfrak{X} \triangledown \mathfrak{G}) \subseteq \mathfrak{X},$$

whence $\mathfrak{X}\mathfrak{G} = \mathfrak{X}$. ■

10.12. PROPOSITION. Let $\mathfrak{Y}\mathfrak{X}_1 = \mathfrak{Y}\mathfrak{X}_2$ where $\mathfrak{X}_1$ and $\mathfrak{X}_2$ are prevarieties and $\mathfrak{Y}$ is a completely decomposable prevariety. Then $\mathfrak{X}_1 = \mathfrak{X}_2$.

P r o o f. Without loss of generality we may assume that $\mathfrak{Y}$ is indecomposable. Suppose $\mathfrak{X}_1 \neq \mathfrak{X}_2$. Let, for instance, $\mathfrak{X}_1 \not\subseteq \mathfrak{X}_2$. Then by Lemma 10.11, $\mathfrak{Y}\mathfrak{G} = \mathfrak{Y}$ which contradicts the assumption. ■

10.13. PROPOSITION. Let $\mathfrak{X}_1\mathfrak{Y} = \mathfrak{X}_2\mathfrak{Y}$ where $\mathfrak{X}_1$ and $\mathfrak{X}_2$ are prevarieties and $\mathfrak{Y}$ is a completely decomposable prevariety. Then $\mathfrak{X}_1 = \mathfrak{X}_2$.

P r o o f. As usual, we may assume $\mathfrak{Y}$ to be indecomposable. Suppose, for instance, that $\mathfrak{X}_1 \not\subseteq \mathfrak{X}_2$. Take $(A, G_1) \in \mathfrak{X}_1 \setminus \mathfrak{X}_2$, $(B, G_2) \in \mathfrak{Y}$ and let $(V,G) = (A, G_1) \triangledown (B, G_2)$. Then $(V, G) \in \mathfrak{X}_1\mathfrak{Y} = \mathfrak{X}_2\mathfrak{Y}$, whence $(\mathfrak{Y}^*(V, G), G) \in \mathfrak{X}_2$. Since $(A, G_1) \notin \mathfrak{X}_2$, we have $\mathfrak{Y}^*(V, G) \subset A$. Denote $A_o = \mathfrak{Y}^*(V, G)$, $\bar{A} = A/A_o$; then

$$(V/A_o, G) = (\bar{A} \oplus B, G) \in \mathfrak{Y}. \tag{10}$$

The canonical epimorphism of (A, G_1) onto $(\bar{A}, G_1)$ can be extended naturally to the epimorphism

$$\mu : (V, G) \longrightarrow (\bar{A}, G_1) \triangledown (B, G_2).$$

It is clear that $A_o \mu = 0$; hence μ can be "passed" through $(V/A_o, G)$, and we obtain an epimorphism

$$\nu : (V/A_o) \longrightarrow (\bar{A}, G_1) \triangledown (B, G_2).$$

Clearly ν is a right epimorphism. By (10), it follows that $(\bar{A}, G_1) \triangledown (B, G_2) \in \mathfrak{Y}$.

Since $\bar{A} \neq 0$ (for $A_o \subset A$), the representation $(\bar{A}, G_1)$ contains the one-dimensional trivial representation $(K, 1)$. By Lemma 3.4 (b), $(K, 1) \triangledown (B, G_2) \in \mathfrak{Y}$. Taking into account that (B, G_2) is an arbitrary representation from $\mathfrak{Y}$, we conclude that

$$\{(K, 1)\} \triangledown \mathfrak{Y} \subseteq \mathfrak{Y}.$$

It is fairly obvious that $\mathfrak{G} = VD\{(K, 1)\}$; hence

$$\mathfrak{G} \triangledown \mathfrak{Y} = (VD\{(K, 1)\}) \triangledown \mathfrak{Y} \subseteq \mathfrak{Y}$$

and, moreover,

$$\mathfrak{G}\mathfrak{Y} \subseteq VSQ_r(\mathfrak{G} \triangledown \mathfrak{Y}) \subseteq \mathfrak{Y}$$

(Lemmas 3.4 and 3.5). Therefore $\mathfrak{G}\mathfrak{Y} = \mathfrak{Y}$ which contradicts the indecomposability of $\mathfrak{Y}$. ∎

10.14. PROPOSITION. Let $\mathfrak{X}_1\mathfrak{X}_2 = \mathfrak{Y}_1\mathfrak{Y}_2$ where $\mathfrak{X}_i$ and $\mathfrak{Y}_i$ are prevarieties, and let $\mathfrak{X}_1$ be completely decomposable. If $\mathfrak{X}_2 \not\subseteq \mathfrak{Y}_2$, then there exists a nontrivial prevariety $\mathfrak{Z}$ such that $\mathfrak{X}_2 = \mathfrak{Z}\mathfrak{Y}_2$.

P r o o f. As in the proof of Lemma 10.11, define

$$\mathfrak{Z}_o = \{(\mathfrak{Y}_2^*(V, G), G) \mid (V, G) \in \mathfrak{X}_2\}.$$

Then $\mathfrak{Z} = \text{pvar}\, \mathfrak{Z}_o \neq \mathfrak{E}$ and, as in Lemma 10.11, it is proved that

$$V\mathfrak{Z}_o = \mathfrak{Z}_o,\ D\mathfrak{Z}_o = \mathfrak{Z}_o,\ \mathfrak{X}_1 \triangledown \mathfrak{Z}_o \subseteq \mathfrak{Y}_1. \tag{11}$$

Furthermore,

$$\mathfrak{X}_2 \subseteq \mathfrak{Z}_o\mathfrak{Y}_2,\ \mathfrak{X}_1\mathfrak{Z}_o \subseteq \mathfrak{Y}_1. \tag{12}$$

Indeed, the first inclusion is trivial, while the second is deduced from (11) and Lemma 3.5 (k):

$$\mathfrak{X}_1\mathfrak{Z}_o \subseteq VSQ_r(\mathfrak{X}_1 \triangledown \mathfrak{Z}_o) \subseteq VSQ_r\mathfrak{Y}_1 = \mathfrak{Y}_1.$$

Since the multiplication of classes of representations is associative, it follows from (12) that $\mathfrak{X}_1\mathfrak{X}_2 \subseteq \mathfrak{X}_1\mathfrak{Z}_o\mathfrak{Y}_2 \subseteq \mathfrak{Y}_1\mathfrak{Y}_2$. By hypothesis, $\mathfrak{X}_1\mathfrak{X}_2 = \mathfrak{Y}_1\mathfrak{Y}_2$ and therefore $\mathfrak{X}_1\mathfrak{X}_2 = \mathfrak{X}_1\mathfrak{Z}_o\mathfrak{Y}_2$.

Denote $\mathfrak{Z}_o\mathfrak{Y}_2 = \mathfrak{L}$. Since $\mathfrak{Z}_o$ and $\mathfrak{Y}_2$ are closed under the operations V and D, so is $\mathfrak{L}$. Assume $\mathfrak{L} \subseteq \mathfrak{X}_2$. Since $\mathfrak{X}_1\mathfrak{L} = \mathfrak{X}_1\mathfrak{X}_2$, it follows from Lemma

10.11 that $\mathfrak{X}_1\mathfrak{S} = \mathfrak{X}_1$. This is impossible because of 10.12, and so $\mathfrak{L} = \mathfrak{Z}_o\mathfrak{Y}_2 \subseteq$ $\subseteq \mathfrak{X}_2$. Consequently,

$$\mathfrak{Z}_o \triangledown \mathfrak{Y}_2 \subseteq (V\mathfrak{Z}_o)(V\mathfrak{Y}_2) = \mathfrak{Z}_o\mathfrak{Y}_2 \subseteq \mathfrak{X}_2$$

and, by Lemma 3.4,

$$\mathfrak{Z} \triangledown \mathfrak{Y}_2 = (VQSC_r\mathfrak{Z}_o) \triangledown \mathfrak{Y}_2 \subseteq \text{pvar}\,(\mathfrak{Z}_o \triangledown \mathfrak{Y}_2) \subseteq \mathfrak{X}_2.$$

Finally,

$$\mathfrak{Z}\mathfrak{Y}_2 \subseteq VSQ_r(\mathfrak{Z} \triangledown \mathfrak{Y}_2) \subseteq \mathfrak{X}_2,$$

but the inclusion $\mathfrak{X}_2 \subseteq \mathfrak{Z}\mathfrak{Y}_2$ follows from (12). Thus $\mathfrak{X}_2 = \mathfrak{Z}\mathfrak{Y}_2$, as required. ∎

P r o o f of Theorem 10.4 is completed now as that of Theorem 10.3. ∎

Thus we have proved the main assertions of the section. Now we want to deduce several corollaries. First we will show that each of Theorems 10.3 and 10.4 is a substantial generalization of Theorem 9.1.

10.15. PROPOSITION. If $\mathfrak{X}$ and $\mathfrak{Y}$ are V-closed classes of representations, then

$$\text{var}\,(\mathfrak{X}\mathfrak{Y}) = \text{var}\,\mathfrak{X} \cdot \text{var}\,\mathfrak{Y}.$$

P r o o f. By Theorem 6.2' and Lemma 3.4 (a),

$$\text{var}\,\mathfrak{X} \cdot \text{var}\,\mathfrak{Y} = \text{var}\,(\mathfrak{X} \triangledown \mathfrak{Y}) \subseteq \text{var}\,(V\mathfrak{X} \cdot V\mathfrak{Y}) = $$
$$= \text{var}\,(\mathfrak{X}\mathfrak{Y}),$$

but the reverse is obvious. ∎

A radical class (prevariety) $\mathfrak{X}$ is called <u>bounded</u> if var $\mathfrak{X}$ differs from $\mathfrak{D}$, the class of all representations.

10.16. COROLLARY. The bounded radicals (prevarieties) are completely decomposable and form a free subsemigroup in $\mathcal{R}$ (in $\mathcal{P}$).

P r o o f. We prove this and several subsequent statements only for radicals since the case of prevarieties is quite analogous. By 10.15, if $\mathfrak{X}$ and $\mathfrak{Y}$ are bounded radicals, then so is $\mathfrak{X}\mathfrak{Y}$. Furthermore, since every proper variety decomposes into a product of a finite number of indecomposable ones, it follows from 10.15 that a bounded radical $\mathfrak{X}$ is completely decomposable, i.e.

$$\mathfrak{X} = \mathfrak{X}_1 \mathfrak{X}_2 \dots \mathfrak{X}_n$$

where $\mathfrak{X}_i$ are indecomposable radicals. Since $\operatorname{var} \mathfrak{X}_i \subseteq \operatorname{var} \mathfrak{X}$, all $\mathfrak{X}_i$ are bounded as well. The rest is evident. ■

10.17. COROLLARY. The product of a finite number of radicals (prevarieties) is a variety if and only if each factor is a variety.

P r o o f. Let $\mathfrak{X}_1, \dots, \mathfrak{X}_n$ be radicals and let $\mathfrak{X} = \mathfrak{X}_1 \dots \mathfrak{X}_n$ be a variety (of course, we assume that $\mathfrak{X} \neq \mathfrak{O}$). For any class $\mathfrak{K}$ denote, for brevity, $\operatorname{var} \mathfrak{K} = \overline{\mathfrak{K}}$. By 10.15, $\mathfrak{X} = \overline{\mathfrak{X}}_1 \dots \overline{\mathfrak{X}}_n$, whence

$$\mathfrak{X}_1 \dots \mathfrak{X}_n = \overline{\mathfrak{X}}_1 \dots \overline{\mathfrak{X}}_n. \tag{13}$$

All the factors in (13) are bounded, and $\mathfrak{X}_i \subseteq \overline{\mathfrak{X}}_i$. Using 10.16, it is easy to conclude that $\mathfrak{X}_i = \overline{\mathfrak{X}}_i$, $i = 1,\dots,n$, i.e. $\mathfrak{X}_i$ are varieties, as required. ■

10.18. COROLLARY. An indecomposable variety is also indecomposable as a radical class and as a prevariety. ■

It follows from this corollary and Theorem 10.3 (or 10.4) that the semigroup of varieties must be free as well. Thus we have obtained Theorem 9.1 again — now as a corollary from 10.3 or 10.4. Our next aim is to show that the semigroups of radicals and prevarieties are considerably more extensive than the semigroup of varieties.

While varieties over a given K form a set, the semigroups of completely decomposable (even bounded) radicals and prevarieties are not sets. This fact will be established in the following way. For an arbitrary variety $\mathfrak{X}$ denote by $\mathcal{R}(\mathfrak{X})$ ($\mathcal{P}(\mathfrak{X})$) the class of all radicals (prevarieties) $\mathfrak{Y}$ such that $\operatorname{var} \mathfrak{Y} = \mathfrak{X}$. We will show that for many $\mathfrak{X}$ the classes $\mathcal{R}(\mathfrak{X})$ and $\mathcal{P}(\mathfrak{X})$ are not sets.

Let $\mathfrak{m}$ be an infinite cardinality. Denote by $\mathfrak{K}_{\mathfrak{m}}$ the class of all representations

(V, G) such that $\dim V < \mathfrak{m}$, and let

$$\mathfrak{R}_{\mathfrak{m}} = \text{rad } \mathfrak{K}_{\mathfrak{m}}, \qquad \mathfrak{P}_{\mathfrak{m}} = \text{pvar } \mathfrak{K}_{\mathfrak{m}}.$$

One can easily verify that:

(i) $(V, G) \in \mathfrak{R}_{\mathfrak{m}}$ iff for every cyclic KG-submodule of V its K-dimension is $< \mathfrak{m}$;

(ii) $(V, G) \in \mathfrak{P}_{\mathfrak{m}}$ iff there is a set of KG-submodules W_i, $i \in I$, of V such that $\cap W_i = 0$ and $\dim_K (V/W_i) < \mathfrak{m}$ for each $i \in I$.

The following lemma clearly demonstrates the duality between radicals and pre-varieties.

10.19. LEMMA. Let $(V, G) = (A, G_1) \nabla (B, G_2)$.

(i) If $\dim A \geqslant \mathfrak{m}$ and $B \neq 0$, then $(V, G) \notin \mathfrak{R}_{\mathfrak{m}}$.

(ii) If $A \neq 0$ and $\dim B \geqslant \mathfrak{m}$, then $(V, G) \notin \mathfrak{P}_{\mathfrak{m}}$

P r o o f. (i) If $0 \neq b \in B$, then the cyclic KG-submodule $b \circ KG$ of V contains A by Corollary 2.10. Hence $\dim (b \circ KG) \geqslant \mathfrak{m}$ and the statement follows.

(ii) Assume the opposite. Then V contains a set of KG-submodules W_i, $i \in I$, such that $\cap W_i = 0$ and $\dim (V/W_i) < \mathfrak{m}$. Consequently, $W_i \cap B \neq 0$ for any $i \in I$. By Corollary 2.10 we have $A \subseteq W_i$, whence $A \subseteq \cap W_i = 0$ which is impossible. ■

10.20. PROPOSITION. If a variety $\mathfrak{X}$ is decomposable, then the classes $\mathcal{R}(\mathfrak{X})$ and $\mathcal{P}(\mathfrak{X})$ are not sets.

P r o o f. Let $\mathfrak{X} = \mathfrak{X}_1 \mathfrak{X}_2$ where $\mathfrak{X}_i$ are nontrivial varieties. For any infinite cardinality $\mathfrak{m}$ choose a free representation $\rho_{i\mathfrak{m}}$ of dimension $\mathfrak{m}$ in $\mathfrak{X}_i$, $i = 1,2$, and let

$$\rho_{\mathfrak{m}} = \rho_{1\mathfrak{m}} \nabla \rho_{2\mathfrak{m}}, \quad \mathfrak{X}_{\mathfrak{m}} = \text{rad } \rho_{\mathfrak{m}}.$$

It follows easily from Theorem 6.2 that $\text{var } \mathfrak{X}_{\mathfrak{m}} = \mathfrak{X}$ for each $\mathfrak{m}$. Therefore $\mathfrak{X}_{\mathfrak{m}} \in \mathcal{R}(\mathfrak{X})$ and it suffices to show that $\mathfrak{m} \neq \mathfrak{n} \implies \mathfrak{X}_{\mathfrak{m}} \neq \mathfrak{X}_{\mathfrak{n}}$.

Let, for example, $\mathfrak{m} < \mathfrak{n}$. By Lemma 10.19 (i), $\rho_{\mathfrak{n}} \notin \mathfrak{R}_{\mathfrak{n}}$. Since the inclusion

$\mathfrak{X}_{\mathfrak{m}} \subseteq \mathfrak{R}_{\mathfrak{n}}$ is obvious, it follows that $\rho_{\mathfrak{n}} \notin \mathfrak{X}_{\mathfrak{m}}$. On the other hand, $\rho_{\mathfrak{n}} \in \mathfrak{X}_{\mathfrak{n}}$ and so $\mathfrak{X}_{\mathfrak{m}} \neq \mathfrak{X}_{\mathfrak{n}}$. ■

In view of Theorems 10.3 and 10.4, indecomposable radicals and prevarieties are of particular interest. In this connection, let us indicate some examples. A radical is called small if it is generated by some set of representations (or equivalently, by a single representation). A small prevariety is defined in a similar way.

10.21. PROPOSITION (Plotkin [50], Vovsi [69]). Small radicals and prevarieties are indecomposable.

Proof. Clearly every small radical $\mathfrak{X}$ is contained in some $\mathfrak{R}_{\mathfrak{m}}$ for a suitable cardinality $\mathfrak{m}$. Suppose that $\mathfrak{X} = \mathfrak{X}_1\mathfrak{X}_2$ where $\mathfrak{X}_i$ are nontrivial radicals. We can find $(A, G_1) \in \mathfrak{X}_1$ and $(B, G_2) \in \mathfrak{X}_2$ such that $\dim A > \mathfrak{m}$ and $B \neq 0$. By Lemma 10.19 (i),

$$(A, G_1) \triangledown (B, G_2) \notin \mathfrak{R}_{\mathfrak{m}};$$

on the other hand,

$$(A, G_1) \triangledown (B, G_2) \in \mathfrak{X}_1\mathfrak{X}_2 = \mathfrak{X}.$$

Contradiction. ■

In order to summarize all of this section, denote by $\mathcal{R}_1$ and $\mathcal{R}_2$ the semigroups of completely decomposable and bounded radicals respectively. Then we have a "descending chain of semigroups"

$$\mathcal{R} \supset \mathcal{R}_1 \supset \mathcal{R}_2 \supset \mathcal{M} \qquad (14)$$

Here $\mathcal{R}_1$, $\mathcal{R}_2$ and $\mathcal{M}$ are free semigroups, but $\mathcal{R}$ is not free — it contains idempotents and has some other "bad" features (this will be discussed a little in the next section). By 10.20 and 10.21, each term of (14) is substantially "greater" than the subsequent one. The same holds for prevarieties.

We remark in conclusion that, as noted in §9, the freeness of the semigroup of varieties can be proved in terms of verbal ideals of the group algebra KF using the technique of Bergman-Lewin. In the investigation of the semigroups of prevarieties and radical classes such an approach is inapplicable. At the present time, the only tool for this purpose is the technique of triangular products.

11. Infinite products of radical classes and prevarieties

An interesting peculiarity of the semigroups $\mathcal{R}$ and $\mathcal{P}$ is that, apart from usual finite products, one can consider the product of an arbitrary totally ordered set of elements in both these semigroups. The definitions are analogous to those of Plotkin [49] concerning radicals and prevarieties of groups.

Let γ be an arbitrary ordinal and let for each nonlimit α, $1 \leqslant \alpha \leqslant \gamma$, there is given a class of representations $\mathfrak{X}_\alpha$. The upper product of these classes is the class $\prod_{\alpha=1}^{\gamma} \mathfrak{X}_\alpha$ of all representations (V, G) such that V possesses an ascending G-invariant series

$$0 = A_0 \subseteq A_1 \subseteq \dots A_\alpha \subseteq A_{\alpha+1} \subseteq \dots A_\gamma = V,$$

where $(A_{\alpha+1}/A_\alpha, G) \in \mathfrak{X}_{\alpha+1}$ for each $\alpha < \gamma$; and $A_\lambda = \bigcup_{\alpha<\lambda} A_\alpha$ for a limit ordinal λ.

The lower product of $\mathfrak{X}_\alpha$, $1 \leqslant \alpha \leqslant \gamma$, is the class $\coprod_{\alpha=1}^{\gamma} \mathfrak{X}_\alpha$ of all representations (V, G) such that V possesses a descending G-invariant series

$$V = A_0 \supseteq A_1 \supseteq \dots A_\alpha \supseteq A_{\alpha+1} \supseteq \dots A_\gamma = 0,$$

where $(A_\alpha/A_{\alpha+1}, G) \in \mathfrak{X}_{\alpha+1}$ for each $\alpha < \gamma$.

The duality between $\mathcal{R}$ and $\mathcal{P}$ results in the following proposition whose proof is straightforward.

11.1. PROPOSITION. The upper product of radicals is a radical. The lower product of prevarieties is a prevariety. ■

In particular, let $\{\mathfrak{X}_\alpha \mid 1 \leqslant \alpha \leqslant \gamma$, α is nonlimit$\}$ be a set of varieties. Then

$\prod\limits_{\alpha=1}^{\gamma} \mathfrak{X}_\alpha$ is a radical, but $\coprod\limits_{\alpha=1}^{\gamma} \mathfrak{X}_\alpha$ is a prevariety. This explains why, dealing with varieties, we naturally come to radicals and prevarieties.

REMARK. When we are concerned with finite products of classes of representations, the difference between the upper product and the lower one is not essential, for the corresponding semigroups are anti-isomorphic. Therefore in the previous section$\frac{2}{3}$ we used the upper product for prevarieties as well as for radicals.

Let $\mathfrak{X}$ be a radical and α an ordinal. According to the above definition, the (upper) power $\mathfrak{X}^\alpha$ is the class of all representations (V, G) such that V possesses an ascending G-series of length $\leq \alpha$ with $\mathfrak{X}$-factors. Clearly $\mathfrak{X} \subseteq \mathfrak{X}^2 \subseteq \mathfrak{X}^3 \subseteq \subseteq \ldots \subseteq \mathfrak{X}^\alpha \subseteq \mathfrak{X}^{\alpha+1} \subseteq \ldots$.

Now let (V, G) be an arbitrary representation. The <u>upper $\mathfrak{X}$-radical series</u> of (V, G) is the series

$$0 = R_0 \subseteq R_1 \subseteq \ldots R_\alpha \subseteq R_{\alpha+1} \subseteq \ldots$$

where $R_1 = \mathfrak{X}(V, G)$, $R_{\alpha+1}/R_\alpha = \mathfrak{X}(V/R_\alpha, G)$, and $R_\lambda = \bigcup_{\alpha<\lambda} R_\alpha$ if λ is a limit ordinal. Evidently the α-th term of this series coincides with the $\mathfrak{X}^\alpha$-radical $\mathfrak{X}^\alpha(V, G)$ of (V, G).

A radical class $\mathfrak{X}$ is called <u>free</u> if all its powers $\mathfrak{X}^\alpha$ are distinct for all ordinals α :

$$\mathfrak{X} \subset \mathfrak{X}^2 \subset \mathfrak{X}^3 \subset \ldots \mathfrak{X}^\alpha \subset \mathfrak{X}^{\alpha+1} \subset \ldots .$$

A quite opposite situation arises when $\mathfrak{X}$ is an idempotent (i.e. $\mathfrak{X} = \mathfrak{X}^2$), for then

$$\mathfrak{X} = \mathfrak{X}^2 = \mathfrak{X}^3 = \ldots = \mathfrak{X}^\alpha = \mathfrak{X}^{\alpha+1} = \ldots .$$

It turns out that the following alternative holds.

11.2. THEOREM. Every radical class is either free or an idempotent.

Proof. Suppose $\mathfrak{X} \neq \mathfrak{X}^2$; we shall prove that $\mathfrak{X}$ is free. It suffices to show that for every ordinal β there exists a representation (V, G) such that

$$(V, G) \in \mathfrak{X}^\beta \quad \text{and} \quad \forall \alpha < \beta : (V, G) \notin \mathfrak{X}^\alpha .$$

technique of Bergman-Lewin. In the investigation of the semigroups of prevarieties and radical classes such an approach is inapplicable. At the present time, the only tool for this purpose is the technique of triangular products.

11. Infinite products of radical classes and prevarieties

An interesting peculiarity of the semigroups $\mathcal{R}$ and $\mathcal{P}$ is that, apart from usual finite products, one can consider the product of an arbitrary totally ordered set of elements in both these semigroups. The definitions are analogous to those of Plotkin [49] concerning radicals and prevarieties of groups.

Let γ be an arbitrary ordinal and let for each nonlimit α, $1 \leqslant \alpha \leqslant \gamma$, there is given a class of representations $\mathfrak{X}_\alpha$. The upper product of these classes is the class $\prod_{\alpha=1}^{\gamma} \mathfrak{X}_\alpha$ of all representations (V, G) such that V possesses an ascending G-invariant series

$$0 = A_0 \subseteq A_1 \subseteq \dots A_\alpha \subseteq A_{\alpha+1} \subseteq \dots A_\gamma = V,$$

where $(A_{\alpha+1}/A_\alpha, G) \in \mathfrak{X}_{\alpha+1}$ for each $\alpha < \gamma$; and $A_\lambda = \bigcup_{\alpha<\lambda} A_\alpha$ for a limit ordinal λ.

The lower product of $\mathfrak{X}_\alpha$, $1 \leqslant \alpha \leqslant \gamma$, is the class $\coprod_{\alpha=1}^{\gamma} \mathfrak{X}_\alpha$ of all representations (V, G) such that V possesses a descending G-invariant series

$$V = A_0 \supseteq A_1 \supseteq \dots A_\alpha \supseteq A_{\alpha+1} \supseteq \dots A_\gamma = 0,$$

where $(A_\alpha/A_{\alpha+1}, G) \in \mathfrak{X}_{\alpha+1}$ for each $\alpha < \gamma$.

The duality between $\mathcal{R}$ and $\mathcal{P}$ results in the following proposition whose proof is straightforward.

11.1.PROPOSITION. The upper product of radicals is a radical. The lower product of prevarieties is a prevariety. ■

In particular, let $\{\mathfrak{X}_\alpha \mid 1 \leqslant \alpha \leqslant \gamma, \alpha \text{ is nonlimit}\}$ be a set of varieties. Then

$\prod_{\alpha=1}^{\gamma} \mathfrak{X}_\alpha$ is a radical, but $\coprod_{\alpha=1}^{\gamma} \mathfrak{X}_\alpha$ is a prevariety. This explains why, dealing with varieties, we naturally come to radicals and prevarieties.

REMARK. When we are concerned with finite products of classes of representations, the difference between the upper product and the lower one is not essential, for the corresponding semigroups are anti-isomorphic. Therefore in the previous section$\frac{2}{3}$ we used the upper product for prevarieties as well as for radicals.

Let $\mathfrak{X}$ be a radical and α an ordinal. According to the above definition, the (upper) power $\mathfrak{X}^\alpha$ is the class of all representations (V, G) such that V possesses an ascending G-series of length $\leqslant \alpha$ with $\mathfrak{X}$-factors. Clearly $\mathfrak{X} \subseteq \mathfrak{X}^2 \subseteq \mathfrak{X}^3 \subseteq \subseteq \ldots \subseteq \mathfrak{X}^\alpha \subseteq \mathfrak{X}^{\alpha+1} \subseteq \ldots$.

Now let (V, G) be an arbitrary representation. The <u>upper $\mathfrak{X}$-radical series</u> of (V, G) is the series

$$0 = R_0 \subseteq R_1 \subseteq \ldots R_\alpha \subseteq R_{\alpha+1} \subseteq \ldots$$

where $R_1 = \mathfrak{X}(V, G)$, $R_{\alpha+1}/R_\alpha = \mathfrak{X}(V/R_\alpha, G)$, and $R_\lambda = \bigcup_{\alpha<\lambda} R_\alpha$ if λ is a limit ordinal. Evidently the α-th term of this series coincides with the $\mathfrak{X}^\alpha$-radical $\mathfrak{X}^\alpha(V, G)$ of (V, G).

A radical class $\mathfrak{X}$ is called <u>free</u> if all its powers $\mathfrak{X}^\alpha$ are distinct for all ordinals α:

$$\mathfrak{X} \subset \mathfrak{X}^2 \subset \mathfrak{X}^3 \subset \ldots \mathfrak{X}^\alpha \subset \mathfrak{X}^{\alpha+1} \subset \ldots .$$

A quite opposite situation arises when $\mathfrak{X}$ is an idempotent (i.e. $\mathfrak{X} = \mathfrak{X}^2$), for then

$$\mathfrak{X} = \mathfrak{X}^2 = \mathfrak{X}^3 = \ldots = \mathfrak{X}^\alpha = \mathfrak{X}^{\alpha+1} = \ldots .$$

It turns out that the following alternative holds.

11.2.THEOREM. Every radical class is either free or an idempotent.

P r o o f. Suppose $\mathfrak{X} \neq \mathfrak{X}^2$; we shall prove that $\mathfrak{X}$ is free. It suffices to show that for every ordinal β there exists a representation (V, G) such that

$$(V, G) \in \mathfrak{X}^\beta \quad \text{and} \quad \forall \alpha < \beta : (V, G) \notin \mathfrak{X}^\alpha.$$

We proceed by induction on β. If $\beta = 2$, the assertion is true by hypothesis. Assume that it holds for every $\beta < \gamma$ where $\gamma > 2$, and let us prove that it holds for γ.

a) γ is a limit ordinal. For every β, $\beta < \gamma$, choose in $\mathfrak{X}^{\beta}$ a representation (V_{β}, G_{β}) such that $(V_{\beta}, G_{\beta}) \notin \mathfrak{X}^{\alpha}$ for each $\alpha < \beta$. Then it is clear that the direct product $(V, G) = \prod_{\beta<\gamma}(V_{\beta}, G_{\beta})$ belongs to $\mathfrak{X}^{\gamma}$, but it does not belong to $\mathfrak{X}^{\beta}$ if $\beta < \gamma$.

b) Both γ and $\gamma-1$ are nonlimit ordinals. Choose $(A, G_1) \in \mathfrak{X}^{\gamma-1} \setminus \mathfrak{X}^{\gamma-2}$ and $(B, G_2) \in \mathfrak{X}^2 \setminus \mathfrak{X}$, and let $(V, G) = (A, G_1) \triangledown (B, G_2)$. Then $(V, G) \in \in \mathfrak{X}^{\gamma-1}\mathfrak{X}^2 = \mathfrak{X}^{\gamma+1}$. Since $(A, G_1) \notin \mathfrak{X}^{\gamma-2}$, it follows that $\mathfrak{X}^{\gamma-2}(A, G_1) = A_o \neq A$. By Lemma 10.5, $\mathfrak{X}^{\gamma-2}(V, G) = A_o$. Consider the representation $(V/A_o, G)$. Since (B, G_2) is a homorphic image of $(V/A_o, G)$ and $(B, G_2) \notin \mathfrak{X}$, we have $(V/A_o, G) \notin \mathfrak{X}$. Consequently, $(V, G) \notin \mathfrak{X}^{\gamma-1}$.

Thus $(V, G) \in \mathfrak{X}^{\gamma+1} \setminus \mathfrak{X}^{\gamma-1}$, whence $\mathfrak{X}^{\gamma-1} \subset \mathfrak{X}^{\gamma+1}$. It follows that $\mathfrak{X}^{\gamma-1} \subset \subset \mathfrak{X}^{\gamma}$, as required.

c) Finally, let γ be a nonlimit ordinal, but $\gamma-1$ a limit one. Choose in $\mathfrak{X}^{\gamma-1}$ a representation (A, G_1) such that $(A, G_1) \notin \mathfrak{X}^{\beta}$ for each $\beta < \gamma-1$, and in $\mathfrak{X}$ any representation (B, G_2) with $B \neq 0$. Let $(V, G) = (A, G_1) \triangledown (B, G_2)$. It is enough to show that $(V, G) \in \mathfrak{X}^{\gamma} \setminus \mathfrak{X}^{\gamma-1}$.

First, $(V, G) \in \mathfrak{X}^{\gamma-1} \cdot \mathfrak{X} = \mathfrak{X}^{\gamma}$. On the other hand, since $\gamma-1$ is a limit ordinal, we have

$$\mathfrak{X}^{\gamma-1}(V, G) = \bigcup_{\beta<\gamma} \mathfrak{X}^{\beta}(V, G).$$

The representation (A, G_1) has been chosen in such a way that $\mathfrak{X}^{\beta}(A, G_1) \subset A$ for every $\beta < \gamma-1$. By Lemma 10.5, $\mathfrak{X}^{\beta}(V, G) \subset A$ for every $\beta < \gamma-1$. Therefore $\mathfrak{X}^{\gamma-1}(V, G) = \bigcup \mathfrak{X}^{\beta}(V, G) \subseteq A$, and so $\mathfrak{X}^{\gamma-1}(V, G)$ is strictly less that V. Hence $(V, G) \notin \mathfrak{X}^{\gamma-1}$, as required. ∎

Theorem 11.2 leads to the following question. Let $\mathfrak{X} = \prod_{\alpha=1}^{\gamma} \mathfrak{X}_{\alpha}$ be an arbitrary product of radicals. Under which conditions $\mathfrak{X}$ is an idempotent ? Clearly if one of the factors, say $\mathfrak{X}_{\varkappa}$, is an idempotent and all the other factors $\mathfrak{X}_{\alpha}$ are contained in $\mathfrak{X}_{\varkappa}$, then $\mathfrak{X} = \mathfrak{X}_{\varkappa}$ and so $\mathfrak{X}$ is an idempotent. Our next objective is to prove that this sufficient condition is also necessary.

11.3. THEOREM. An arbitrary product $\prod_{\alpha=1}^{\gamma} \mathfrak{X}_{\alpha}$ of radicals is an idempotent if and only if some factor $\mathfrak{X}_{\varkappa}$ is an idempotent and contains all the other factors.

11.4. COROLLARY. Free radicals form a subsemigroup in the semigroup of radicals. This subsemigroup is closed with respect to forming infinite products. ■

The proof of the theorem depends on two easy lemmas.

10.5. LEMMA. Let $\mathfrak{X}$ and $\mathfrak{Y}$ be radicals such that $\mathfrak{X} \not\subseteq \mathfrak{Y}$ and $\mathfrak{Y} \not\subseteq \mathfrak{X}$. Then:

(i) $\mathfrak{X}\mathfrak{Y} \not\subseteq \mathfrak{Y}\mathfrak{X}$ and $\mathfrak{Y}\mathfrak{X} \not\subseteq \mathfrak{X}\mathfrak{Y}$,

(ii) $\mathfrak{X}\mathfrak{Y}$ and $\mathfrak{Y}\mathfrak{X}$ are free radicals.

Proof. (i) Let $(A, G_1) \in \mathfrak{X} \setminus \mathfrak{Y}$, $(B, G_2) \in \mathfrak{Y} \setminus \mathfrak{X}$. Then

$$(V, G) = (A, G_1) \triangledown (B, G_2) \in \mathfrak{X}\mathfrak{Y}.$$

Since $\mathfrak{Y}(A, G_1) \subset A$, we obtain by Lemma 10.5 that $W = \mathfrak{Y}(V, G) = \mathfrak{Y}(A, G_1)$. Therefore

$$(V/W, G) \cong (A/W, G_1) \triangledown (B, G_2) \notin \mathfrak{X}$$

since $(B, G_2) \in \mathfrak{X}$. Consequently, $(V, G) \notin \mathfrak{Y}\mathfrak{X}$, whence $\mathfrak{X}\mathfrak{Y} \not\subseteq \mathfrak{Y}\mathfrak{X}$. Analogously $\mathfrak{Y}\mathfrak{X} \not\subseteq \mathfrak{X}\mathfrak{Y}$.

(ii) Suppose that $\mathfrak{X}\mathfrak{Y}$ is not free. By Theorem 11.2, $\mathfrak{X}\mathfrak{Y}$ is an idempotent, whence

$$\mathfrak{X}\mathfrak{Y}\mathfrak{X} = \mathfrak{X}\mathfrak{Y}.$$

This is impossible, since we have just proved that even $\mathfrak{Y}\mathfrak{X}$ is not contained in $\mathfrak{X}\mathfrak{Y}$. ■

11.6. LEMMA. Let $\mathfrak{X}$ and $\mathfrak{Y}$ be radicals such that $\mathfrak{X} \subseteq \mathfrak{Y} \subseteq \mathfrak{X}^2$. If $\mathfrak{X}$ is free, $\mathfrak{Y}$ is free as well.

$\mathfrak{X}$ is free, $\mathfrak{Y}$ is free as well.

P r o o f. Since $\mathfrak{X} \subseteq \mathfrak{Y} \Rightarrow \mathfrak{X}^3 \subseteq \mathfrak{Y}^3$, we have the following chain of inclusions: $\mathfrak{Y} \subseteq \mathfrak{X}^2 \subset \mathfrak{X}^3 \subseteq \mathfrak{Y}^3$. Thus $\mathfrak{Y} \subset \mathfrak{Y}^3$ so that $\mathfrak{Y}$ cannot be an idempotent. ∎

P r o o f of Theorem 11.3. The sufficiency is trivial. To prove the necessity it is enough to show that if $\mathfrak{X} = \prod_{\alpha=1}^{\gamma} \mathfrak{X}_\alpha$ is an idempotent, then $\mathfrak{X} = \mathfrak{X}_\varkappa$ for some nonlimit $\varkappa < \gamma$. We shall prove this by induction on γ.

a) Let $\gamma = 2$. Then $\mathfrak{X} = \mathfrak{X}_1 \mathfrak{X}_2$ and $\mathfrak{X}$ is an idempotent. By Lemma 11.5, either $\mathfrak{X}_1 \subseteq \mathfrak{X}_2$ or $\mathfrak{X}_2 \subseteq \mathfrak{X}_1$. Assume the first. Then $\mathfrak{X}_2 \subseteq \mathfrak{X} \subseteq \mathfrak{X}_2^2$. If $\mathfrak{X}_2$ is free, we obtain by Lemma 11.6 that $\mathfrak{X}$ is also free, which contradicts the hypothesis. Hence $\mathfrak{X}_2$ is an idempotent and it follows from $\mathfrak{X}_1 \subseteq \mathfrak{X}_2$ that $\mathfrak{X} = \mathfrak{X}_2$.

If $\mathfrak{X}_2 \subseteq \mathfrak{X}_1$, a similar argument shows that $\mathfrak{X} = \mathfrak{X}_1$. Thus the case $\gamma = 2$ is completed.

b) Suppose that for all $\beta < \gamma$ the theorem has already been proved. In order to prove it for γ, suppose first that γ is a nonlimit ordinal. Then

$$\mathfrak{X} = \prod_{\alpha=1}^{\gamma} \mathfrak{X}_\alpha = \left(\prod_{\alpha=1}^{\gamma-1} \mathfrak{X}_\alpha\right) \mathfrak{X}_\gamma$$

and, by a), either $\mathfrak{X} = \mathfrak{X}_\gamma$ or $\mathfrak{X} = \prod_{\alpha=1}^{\gamma-1} \mathfrak{X}_\alpha$. If $\mathfrak{X} = \mathfrak{X}_\gamma$, it is nothing to prove. If $\mathfrak{X} = \prod_{\alpha=1}^{\gamma-1} \mathfrak{X}_\alpha$, it follows from induction hypothesis that $\mathfrak{X} = \mathfrak{X}_\varkappa$ for some $\varkappa \leq \gamma - 1$, as required.

Now let γ be a limit ordinal. For every $\beta < \gamma$ denote $\overline{\mathfrak{X}}_\beta = \prod_{\alpha=1}^{\beta} \mathfrak{X}_\alpha$, then

$$\overline{\mathfrak{X}}_1 \subseteq \overline{\mathfrak{X}}_2 \subseteq \dots \overline{\mathfrak{X}}_\beta \subseteq \overline{\mathfrak{X}}_{\beta+1} \subseteq \dots \subseteq \mathfrak{X}.$$

Suppose $\mathfrak{X} = \overline{\mathfrak{X}}_\beta$ for some $\beta < \gamma$. Then $\overline{\mathfrak{X}}_\beta$ is an idempotent and the theorem follows by induction hypothesis. Suppose that $\mathfrak{X} \neq \overline{\mathfrak{X}}_\beta$ for each $\beta < \gamma$. Then for each $\beta < \gamma$ we can choose a representation $(A_\beta, G_\beta) \in \mathfrak{X} \setminus \overline{\mathfrak{X}}_\beta$. Clearly the representation $(A, G) = \prod_{\beta<\gamma} (A_\beta, G_\beta)$ belongs to $\mathfrak{X} \setminus \bigcup_{\beta<\gamma} \overline{\mathfrak{X}}_\beta$. Let (B, H) be any nonzero representation from $\mathfrak{X}$. Since $\mathfrak{X}$ is an idempotent, it follows that

$$(V, T) = (A, G) \triangledown (B, H) \in \mathfrak{X}.$$

On the other hand, using Lemma 10.5, we obtain

$$(V, T) = \bigcup_{\beta<\gamma} \overline{\mathfrak{X}}_\beta (V, T) = \bigcup_{\beta<\gamma} \overline{\mathfrak{X}}_\beta (A, G) \subseteq A \subset V,$$

which is impossible. ■

Of course, similar results are valid for prevarieties. Having defined in a natural way the notion of a free prevariety, we can establish the following theorems; their proofs are dual to those of Theorems 11.2 and 11.3 and are left to the reader.

11.7.THEOREM. Every prevariety is either free or an idempotent.

11.8.THEOREM. An arbitrary product $\bigsqcup_{\alpha=1}^{\gamma} \mathfrak{X}_\alpha$ of prevarieties is an idempotent if and only if some factor $\mathfrak{X}_\varkappa$ is an idempotent and contains all the other factors.

Note that Theorems 11.2, 11.3, 11.7, 11.8 and their proofs are absolutely analogous to the earlier results of the author [67 , 68] concerning radicals and prevarieties of groups. Moreover, all the known results on infinite products of radical classes and prevarieties of groups can be transferred verbatim to the situation of group representations; one should only replace wreath products of groups by triangular products of representations.

However, the technique of triangular products allows to obtain essentially new results. Naturally, these results are of greater interest for us. Some of them will be established here.

As before, $\mathcal{R}$ denotes the semigroup of all radicals (over a given field K) and $\mathcal{R}_1$ its subsemigroup of completely decomposable radicals. Let $\mathcal{S} = \mathcal{R} \setminus \mathcal{R}_1$. The following two theorems have been proved in [75] .

11.9.THEOREM. $\mathcal{S}$ is an ideal of $\mathcal{R}$.

P r o o f. We must prove that if $\mathfrak{X} \in \mathcal{S}$ and $\mathfrak{Y} \in \mathcal{R}$, then both $\mathfrak{X}\mathfrak{Y}$ and $\mathfrak{Y}\mathfrak{X}$ belong to $\mathcal{S}$. We will show, by induction on n, that neither $\mathfrak{X}\mathfrak{Y}$ nor $\mathfrak{Y}\mathfrak{X}$ can decompose into a product of n indecomposable radicals.

For n = 1 this assertion is trivial. Suppose that it holds for n - 1 and let

$$\mathfrak{X}\mathfrak{Y} = \mathfrak{L}_1\mathfrak{L}_2 \dots \mathfrak{L}_n$$

where all $\mathfrak{L}_i$ are indecomposable radicals. Denote $\mathfrak{L}_2 \dots \mathfrak{L}_n = \overline{\mathfrak{L}}$ and apply Proposition 10.9 to the equality $\mathfrak{X}\mathfrak{Y} = \mathfrak{L}_1\overline{\mathfrak{L}}$. If $\mathfrak{L}_1 \nsubseteq \mathfrak{X}$, we obtain that there exists $\mathfrak{Z} \neq \mathfrak{E}$ such that $\mathfrak{L}_1 = \mathfrak{X}\mathfrak{Z}$. This contradics the indecomposability of $\mathfrak{L}_1$, whence $\mathfrak{L}_1 \subseteq \mathfrak{X}$.

Furthermore, if $\mathfrak{X} \nsubseteq \mathfrak{L}_1$, Proposition 10.9 implies that there exists $\mathfrak{Z} \neq \mathfrak{E}$ such that

$$\mathfrak{X}\mathfrak{Y} = \mathfrak{L}_1\mathfrak{Z}\mathfrak{Y} = \mathfrak{L}_1\overline{\mathfrak{L}} \tag{1}$$

and, since $\mathfrak{L}_1$ is indecomposable,

$$\mathfrak{Z}\mathfrak{Y} = \overline{\mathfrak{L}} = \mathfrak{L}_2\mathfrak{L}_3 \dots \mathfrak{L}_n. \tag{2}$$

Consider two cases.

a) $\mathfrak{Y}$ is completely decomposable. By (1) and 10.9, $\mathfrak{X} = \mathfrak{L}_1\mathfrak{Z}$ and so $\mathfrak{X} \in \mathfrak{S} \Rightarrow \mathfrak{Z} \in \mathfrak{S}$. But then (2) is impossible because of induction hypothesis.

b) $\mathfrak{Y}$ is not completely decomposable. Then (2) is not impossible by the same reason.

Thus $\mathfrak{X} = \mathfrak{L}_1$. But this is also impossible, since $\mathfrak{L}_1$ is indecomposable and $\mathfrak{X} \in \mathfrak{S}$. Consequently, $\mathfrak{X}\mathfrak{Y}$ cannot decompose into a product of n indecomposable radicals. Neither can $\mathfrak{Y}\mathfrak{X}$ (the proof is analogous). ■

11.10.THEOREM. A product of an infinite number of nontrivial radicals is not a completely decomposable radical.

P r o o f. Consider a product $\prod_{\alpha=1}^{\gamma} \mathfrak{X}_\alpha$ where γ is infinite and $\mathfrak{X}_\alpha \neq \mathfrak{E}$ for all α. Suppose $\prod_{\alpha=1}^{\gamma} \mathfrak{X}_\alpha = \mathfrak{L}_1\mathfrak{L}_2 \dots \mathfrak{L}_n$ where $\mathfrak{L}_i$ are indecomposable. Denote $\prod_{\alpha=n+1}^{\gamma} \mathfrak{X}_\alpha = \overline{\mathfrak{X}}$, then

$$\mathfrak{X}_1\mathfrak{X}_2 \dots \mathfrak{X}_n\overline{\mathfrak{X}} = \mathfrak{L}_1\mathfrak{L}_2 \dots \mathfrak{L}_n.$$

By Theorem 11.9, all the radicals $\mathfrak{X}_1, \dots, \mathfrak{X}_n, \overline{\mathfrak{X}}$ are completely decomposable. But it follows from Theorem 10.3 that a product of n + 1 completely decomposable radicals cannot be equal to a product of n indecomposable ones. ■

REMARK. It is still unknown whether the similar assertion holds for radical classes of groups; for some partial results see [68].

Our next result deals specifically with varieties of representations. Since such a variety $\mathfrak{X}$ is also a radical class, one can construct in an arbitrary representation both the upper $\mathfrak{X}$-radical series and the lower $\mathfrak{X}$-verbal series. By Theorem 11.2, for any ordinal α there exists a representation (V, G) in which the upper $\mathfrak{X}$-series reaches V in precisely α steps. On the other hand, according to Theorem 11.7, for any ordinal β there exists a representation (W, H) in which the lower $\mathfrak{X}$-series reaches zero in precisely β steps. It turns out that both these facts can be joined as follows.

11.11. THEOREM (Vovsi [73]). For any variety of representations $\mathfrak{X}$ and any infinite ordinals α and β there exists a representation (V, G) in which the upper $\mathfrak{X}$-series reaches V in precisely α steps and the lower $\mathfrak{X}$-series reaches zero in precisely β steps.

Sketch of the proof. We use definitions and notation from Section 4. Let Λ be the set of pairs (λ, μ), where λ and μ are nonlimit ordinals satisfying the conditions $1 \leq \lambda \leq \alpha$ and $1 \leq \mu \leq \beta$. Define an order on Λ by the rule

$$(\lambda_1, \mu_1) < (\lambda_2, \mu_2) \iff (\lambda_1 < \lambda_2) \ \& \ (\mu_1 > \mu_2).$$

Then Λ becomes a partially ordered set. It turns out that if $\varrho = \mathrm{Fr}\,\mathfrak{X}$, then the triangular Λ-power $\nabla \varrho^{\Lambda}$ satisfies all the requirements. For the details see [73].

Clearly the condition " α and β are infinite" cannot be omitted. For if the upper $\mathfrak{X}$-series reaches V in n steps (where $n < \infty$), then an obvious argument shows that the lower $\mathfrak{X}$-series reaches zero also in n steps, and vice versa.

In connection with Theorem 11.11, we note one more fact concerning radicals and verbals in infinite triangular products.

11.12. PROPOSITION. Let $(V, G) = \nabla_{\alpha \in \Lambda} (A_\alpha, G_\alpha)$ be a triangular product of nonzero representations and let $\mathfrak{X}$ be a nontrivial variety of representations.

(i) If Λ has no minimal elements then $\mathfrak{X}(V, G) = 0$.

(ii) If Λ has no maximal elements then $\mathfrak{X}^*(V, G) = V$.

For the proof we refer to [73].

12. Invariant subspaces in representations

An important property of a triangular product is that one can describe all its invariant subspaces provided one has such a description for all factors - see Proposition 4.5 and, for finite products, Proposition 2.9. This allows to construct, by means of triangular products, various examples of representations whose lattices of invariant subspaces have certain prescribed properties.

More exactly, for an arbitrary representation (V, G) denote by $L(V, G)$ the lattice of all G-invariant subspaces (G-submodules) of V. There naturally arises a question: which lattices can be realized as $L(V, G)$ for a suitable representation (V, G) ? We will prove the following.

12.1.THEOREM (Vovsi [76]). For every partially ordered set Λ there exists a representation (V, G) such that $L(V, G) \cong 2^{\Lambda}$.

Recall that the lattice 2^{Λ} can be, for example, defined as follows: it is the lattice of all lower segments of Λ (see Birkhoff [4]).

We note that Theorem 12.1 is parallel to that of Silcock [63] concerning the lattices of normal subgroups of groups. The proof of Silcock's theorem is based on wreath products of groups and is substantially more complicated.

P r o o f. Take any non-unit irreducible representation (A, H) and for every $\alpha \in \Lambda$ denote by (A_α, H_α) an isomorphic copy of (A, H). Let

$$V = \underset{\alpha\in\Lambda}{\nabla}(A_\alpha, H_\alpha) = (\underset{\alpha\in\Lambda}{\oplus} A_\alpha, \ \Phi \leftthreetimes \prod_{\alpha\in\Lambda} H_\alpha).$$

For every $\alpha \in \Lambda$ denote by π_α the natural projection of $V = \oplus A_\alpha$ onto A_α. Take any G-submodule W of V and consider in Λ the subset

$$\Omega = \{\alpha \mid W\pi_\alpha \neq 0\}.$$

By Corollary 4.4, Ω is a lower segment of Λ. We will prove that $W = \bigoplus_{\alpha \in \Omega} A_\alpha$. It suffices to show that $\alpha \in \Omega \Rightarrow A_\alpha \subseteq W$.

If $\alpha \in \Omega$, there exists $w \in W$ such that

$$w = a_\alpha + a_{\beta_1} + \dots + a_{\beta_n},$$

where $0 \neq a_\alpha \in A_\alpha$, $a_{\beta_i} \in A_{\beta_i}$. Since (A, H) is a non-unit representation, we can find an element $h_\alpha \in H_\alpha$ such that $a_\alpha \circ h_\alpha \neq a_\alpha$. But $\prod H_\alpha$ acts on $\oplus A_\alpha$ componentwise; hence

$$w \circ h_\alpha = a_\alpha \circ h_\alpha + a_{\beta_1} + \dots + a_{\beta_n}$$

and $w \circ h_\alpha - w = a_\alpha \circ h_\alpha - a_\alpha \neq 0$. Since W is invariant under G, the element $w \circ h_\alpha - w$ belongs to W; so W contains a nonzero element $a_\alpha \circ h_\alpha - a_\alpha$ of the irreducible H_α-module A_α. Therefore $A_\alpha \subseteq W$, as required.

Thus we have proved that if W is any G-submodule of V then $W = \bigoplus_{\alpha \in \Omega} A_\alpha$ for some lower segment Ω of Λ. It follows that the lattice L(V, G) is isomorphic to the lattice of all lower segments of Λ, i.e. the lattice 2^Λ. ∎

Let L be a finite distributive lattice. It is well known that $L \cong 2^\Lambda$ for some finite partially ordered set Λ. Applying the proof of Theorem 12.1 to this Λ and some finite-dimensional irreducible representation (A, H), we obtain

12.2.COROLLARY. For every finite distributive lattice L there exists a finite-dimensional representation (V, G) such that $L(V, G) \cong L$. ∎

We note that the corollary cannot be extended to arbitrary finite modular lattices. This follows from results of Jonsson [23].

Our next objective is to introduce a construction which is to some extent a synthesis of triangular products of representations and wreath products of groups. This construction, in particular, gives a simple method of constructing infinite-dimensional irreducible representations.

The ingredients for this construction are a representation $\varrho = (A, H)$, a partially ordered set Λ, and a group T of automorphisms (= order-preserving bijections) of Λ. Consider the restricted triangular Λ-power of (A, H)

$$(V, G) = \nabla(A, H)^{\Lambda} = (A^{(\Lambda)}, \Phi \lambda H^{(\Lambda)}).$$

The group T acts on $A^{(\Lambda)}$ and $H^{(\Lambda)}$ in the usual way. Namely, if $\bar{a} \in A^{(\Lambda)}$, $\bar{h} \in H^{(\Lambda)}$, $t \in T$, then for every $\alpha \in \Lambda$

$$(\bar{a} \circ t)(\alpha) = \bar{a}(\alpha t^{-1}), \quad (\bar{h} \circ t)(\alpha) = \bar{h}(\alpha t^{-1}).$$

Evidently both these actions are faithful. Therefore the group T may be regarded as a subgroup of Aut V. The group Φ is also contained in Aut V, and we claim that

(i) $\Phi \lhd \langle \Phi, T \rangle$,

(ii) $\Phi \cap T = 1$.

To prove (i), let $\varphi \in \Phi$, $t \in T$. We have to show that $t^{-1}\varphi t \in \Phi$. From now on, we shall denote the direct summand (factor) in $A^{(\Lambda)}$ (in $H^{(\Lambda)}$) which corresponds to an arbitrary $\alpha \in \Lambda$ by A_α (H_α). We will show that for any $a \in A_\alpha$

$$a \circ (t^{-1}\varphi\, t) = a + b \quad \text{where } b \in V_\alpha^\flat = \bigoplus_{\lambda < \alpha} A \quad . \tag{1}$$

This will imply that $t^{-1}\varphi\, t$ stabilizes the system of factors $\{V_\alpha / V_\alpha^\flat\}$, i.e. $t^{-1}\varphi\, t \in \Phi$. Since $a \in A_\alpha$, we have $a \circ t^{-1} \in A_{\alpha t^{-1}}$. Therefore

$$(a \circ t^{-1}) \circ \varphi = a \circ t^{-1} + b_1 \quad \text{where } b_1 \in V^\flat_{\alpha t^{-1}}.$$

Consequently $a \circ (t^{-1}\varphi\, t) = a + b_1 \circ t$. Since t is an automorphism of Λ, it is easy to see that $b_1 \circ t \in V^\flat_{\alpha t^{-1} t} = V^\flat_\alpha$. Denoting $b_1 \circ t = b$, we obtain (1).

(ii) Suppose $t \in \Phi \cap T$. Then for an arbitrary $a \in A_\alpha$

$$a \circ t \in A_{\alpha t}. \tag{2}$$

Since $t \in \Phi$, we have

$$a \circ t = a + b \quad \text{where } b \in V \quad . \tag{3}$$

Comparing (2) and (3), we conclude that $a \circ t = a$. Hence $t = 1$, as claimed.

So the subgroup $\langle \Phi, T \rangle$ of Aut V splits in a semidirect product $\Phi \lambda T$. In particular, T acts on Φ by conjugations. This action we shall also denote by $\circ$, that is, if $\varphi \in \Phi$ and $t \in T$ then $\varphi \circ t \overset{\text{def.}}{=} t^{-1}\varphi\, t$.

Thus the group T acts on Φ and on $H^{(\Lambda)}$. The crucial question now is whether these two actions give an action of T on $G = \Phi \curlywedge H^{(\Lambda)}$. To answer this question in the affirmative, we must establish that for arbitrary $\varphi \in \Phi$, $\bar{h} \in H^{(\Lambda)}$ and $t \in T$ the following equality holds:

$$(\bar{h}^{-1}\varphi\, \bar{h})\circ t = (\bar{h}\circ t)^{-1}(\varphi\circ t)(\bar{h}\circ t). \tag{4}$$

Both parts here are elements of Φ. Since Φ acts faithfully on V, it is enough to show that both parts of (4) act equally on V.

For any $a \in A$ and $\alpha \in \Lambda$ denote by a_α the element-function from $A^{(\Lambda)}$ defined as follows

$$a_\alpha(\alpha) = a \quad \text{and} \quad a_\alpha(\lambda) = 0 \quad \text{if } \lambda \neq \alpha .$$

Clearly $a_\alpha \in A_\alpha$ and V is generated by all such a_α. Therefore it is sufficient to show that for every a_α

$$a_\alpha \circ ((\bar{h}^{-1}\varphi\, \bar{h})\circ t) = a_\alpha \circ ((\bar{h}\circ t)^{-1}(\varphi\circ t)(\bar{h}\circ t)). \tag{5}$$

Since $a_{\alpha t^{-1}} \in A_{\alpha t^{-1}}$, it follows that $a_{\alpha t^{-1}} \circ \bar{h}^{-1} \in A_{\alpha t^{-1}}$ as well, whence

$$(a_{\alpha t^{-1}} \circ \bar{h}^{-1})\circ \varphi = a_{\alpha t^{-1}} \circ \bar{h}^{-1} + b \tag{6}$$

where $b \in V^{b}_{\alpha t^{-1}}$. Using the definitions of all the actions introduced above and (6), we obtain

1)
$$\begin{aligned} a_\alpha \circ ((\bar{h}^{-1}\varphi\, \bar{h})\circ t) &= ((a_\alpha \circ t^{-1})\circ(\bar{h}^{-1}\varphi\, \bar{h}))\circ t = \\ &= (((a_{\alpha t^{-1}} \circ \bar{h}^{-1})\circ \varphi)\circ \bar{h})\circ t = ((a_{\alpha t^{-1}} \circ \bar{h}^{-1} + b)\circ \bar{h})\circ t = \\ &= (a_{\alpha t^{-1}} + b\circ \bar{h})\circ t = a_\alpha + (b\circ\bar{h})\circ t; \end{aligned} \tag{7}$$

2)
$$\begin{aligned} a_\alpha \circ ((\bar{h}\circ t)^{-1}(\varphi\circ t)(\bar{h}\circ t)) &= ((a_\alpha \circ (\bar{h}\circ t)^{-1})\circ t^{-1}\varphi\, t)\circ(\bar{h}\circ t) = \\ &= ((a\circ \bar{h}(\alpha t^{-1})^{-1})_\alpha \circ t^{-1}\varphi\, t)\circ(\bar{h}\circ t) = ((a\circ\bar{h}(\alpha t^{-1})^{-1})_{\alpha t^{-1}} \circ \varphi t)\circ(\bar{h}\circ t) = \\ &= (a_{\alpha t^{-1}} \circ \bar{h}^{-1})\circ \varphi t)\circ(\bar{h}\circ t) = ((a_{\alpha t^{-1}} \circ \bar{h}^{-1})\circ t + b\circ t)\circ(\bar{h}\circ t) = \\ &= ((a_{\alpha t^{-1}} \circ \bar{h}^{-1})\circ t)\circ(\bar{h}\circ t) + (b\circ t)\circ(\bar{h}\circ t). \end{aligned} \tag{8}$$

It is easy to see that for any $\bar{a} \in V$, $\bar{h} \in H^{(\Lambda)}$, $t \in T$

$$(\bar{a}\circ t)\circ(\bar{h}\circ t) = (\bar{a}\circ\bar{h})\circ t. \tag{9}$$

Therefore

$$((a_{\alpha t^{-1}}\circ \bar{h}^{-1})\circ t)\circ(\bar{h}\circ t) = a_{\alpha t^{-1}}\circ t = a_{\alpha},$$
$$(b\circ t)\circ(\bar{h}\circ t) = (b\circ\bar{h})\circ t. \tag{10}$$

By (8) and (10),

$$a_{\alpha}\circ((\bar{h}\circ t^{-1})(\varphi\circ t)(\bar{h}\circ t)) = a_{\alpha} + (b\circ\bar{h})\circ t. \tag{11}$$

The required equality (5) follows now from (7) and (11).

Thus the group T acts on the group $\Phi \leftthreetimes H^{(\Lambda)}$. Hence there arises a semidirect product $(\Phi \leftthreetimes H^{(\Lambda)}) \leftthreetimes T = \Phi \leftthreetimes H^{(\Lambda)} \leftthreetimes T$. All the factors Φ, $H^{(\Lambda)}$ and T of this semidirect product act on $V = A^{(\Lambda)}$. These three actions are "well-agreed": the actions of Φ and $H^{(\Lambda)}$ agree by the definition of triangular product, the actions of Φ and T agree since $\Phi \leftthreetimes T$ is regarded as a subgroup of Aut V, and the actions of $H^{(\Lambda)}$ and T agree in view of (9). These observations show that we obtained a representation

$$(V,\ \Phi \leftthreetimes H^{(\Lambda)} \leftthreetimes T).$$

It is called the <u>wreathed triangular Λ-power of (A, H) by T</u> and is denoted by

$$\nabla(A, H)^{\Lambda} \text{ wr } T.$$

An important special case should be mentioned. Let Λ be an arbitrary set. We may assume that Λ is partially ordered under the trivial ordering $\lambda \leq \mu \iff \lambda = \mu$. Then it is evident that $\nabla(A, H)^{\Lambda}$ degenerates into a direct power $(A, H)^{(\Lambda)} = (A^{(\Lambda)}, H^{(\Lambda)})$.

If T is any group acting on Λ, the corresponding wreathed triangular power has the following form:

$$\nabla(A, H)^{\Lambda} \text{ wr } T = (A^{(\Lambda)}, H^{(\Lambda)} \leftthreetimes T). \tag{12}$$

The reader familiar with the concept of wreath product can easily realize that the group $H^{(\Lambda)} \leftthreetimes T$ in (12) is the wreath product $H \overset{\Lambda}{\text{wr}} T$ of the groups H and T over the set Λ. The representation (12) is called the <u>wreath product of (A, H) and T over Λ</u>

and is also denoted by

$$(A, H) \overset{\Lambda}{\mathrm{wr}}\ T = (A^{(\Lambda)}, H \overset{\Lambda}{\mathrm{wr}}\ T).$$

In particular, if $\Lambda = T$ and T acts on itself by right multiplication, we obtain the _standard_ wreath product of (A, H) and T

$$(A, H)\ \mathrm{wr}\ T = (A^{(T)}, H\ \mathrm{wr}\ T)$$

in the sense of Plotkin [52].

REMARK. All the constructions considered here are _restricted_, i.e. they are based on restricted direct powers of (A, H). Slightly modifying our argument, one can easily define the corresponding _complete_ constructions which are based on Cartesian powers. The necessary alterations are straightforward.

We will indicate now some applications of wreathed triangular powers. Let Λ be a linearly ordered set and T a group of automorphisms of Λ. Following P.Hall [18] we say that T acts on Λ _irreducibly_ if

$$(\forall \lambda, \mu \in \Lambda)(\exists t \in T)[\lambda t > \mu].$$

Evidently in such a case Λ possesses neither maximal, nor minimal elements.

12.3.THEOREM (Vovsi [73]). Let (A, H) be a representation, Λ a lineraly ordered set, and T a group acting irreducibly on Λ. Then

$$\nabla (A, H)^{\Lambda}\ \mathrm{wr}\ T$$

is an irreducible representation.

P r o o f. Denote

$$(V, G) = \nabla (A, H)^{\Lambda}\ \mathrm{wr}\ T = (A^{(\Lambda)}, \Phi\lambda H^{(\Lambda)} \lambda\ T)$$

and let W be a nonzero G-submodule of V. In particular, W is invariant under $\Phi\lambda H^{(\Lambda)}$. Since Λ is lineraly ordered, it follows from Proposition 4.5 that either $W = V_{\Omega}$ for some lower segment Ω of Λ, or $W = V_{\alpha}^{b} \oplus B$ where $\alpha \in \Lambda$ and B is a nonzero H_{α}-submodule of A_{α}.

First let $W = V_{\Omega}$. Suppose that $\Omega \neq \Lambda$ and choose $\lambda \in \Lambda \setminus \Omega$. Take an arbitrary nonzero element of W

$$w = a_1 + a_2 + \dots + a_n \quad (a_i \in A_{\alpha_i}, \ \alpha_1 < \dots < \alpha_n)$$

and find $t \in T$ such that $\alpha_n t > \lambda$. Clearly $w \cdot t \notin V_\Omega$ but, on the other hand, $w \cdot t \in$ $\in W$ since W is invariant under G. This contradiction shows that $\Omega = \Lambda$, whence $W = V$.

If $W = V_\alpha^b \oplus B$, it is sufficient to take α instead of λ and repeat the previous argument. ■

Using this theorem one can construct various examples of infinite-dimensional irreducible representations. For example, let (A, H) be an arbitrary representation and $\mathbb{Z}$ the set of integers in their natural order. Take the transformation $t : n \mapsto n + 1$ of $\mathbb{Z}$ and consider the representation

$$\nabla (A, H)^{\mathbb{Z}} \text{ wr } \langle t \rangle .$$

By Theorem 12.3 it is irreducible.

We conclude this section by outlining another application of our construction. A well known theorem of P.Hall [19] states that there exist non-strictly simple groups, that is, simple groups which possess ascending normal series of non-unit length. A parallel result for representations is valid as well. Indeed, let (V, G) be a representation. Among all normal representations of (V, G) we choose the following:

a) normal subrepresentations of the kind (V, H) where $H \lhd G$;

b) normal subrepresentations of the kind $(0, H)$ where $\operatorname{Ker}(V, G) \supseteq H \lhd G$.

It is natural to call them <u>trivial</u>. In other words, a normal subrepresentation (A, H) of (V, G) is nontrivial if $0 \neq A \neq V$. Evidently a representation has no nontrivial normal subrepresentations if and only if it is irreducible.

An ascending normal series of a representation (V, G) is a series

$$(0, 1) = (A_0, H_0) \lhd (A_1, H_1) \lhd \dots (A_\gamma, H_\gamma) = (V, G) \qquad (13)$$

such that (A_α, H_α) is a nontrivial normal subrepresentation of (A_{+1}, H_{+1}), and $(A_\lambda, H_\lambda) = \bigcup_{\alpha < \lambda} (A_\alpha, H_\alpha)$ if λ is a limit ordinal. The ordinal γ is called the <u>length</u>

of the series (13).

DEFINITION. A representation is said to be strictly irreducible if it has no ascending normal series of length > 1.

Clearly a strictly irreducible representation is irreducible. The following theorem asserts that the reverse is false. Its proof is based on ideas of P.Hall [19].

12.4.THEOREM. For every limit ordinal λ there exists an irreducible representation which has an ascending normal series of length $\omega\lambda$.

Sketch of the proof. Let $\varrho = (A, H)$ be a nonzero representation, $\mathbb{Z}$ the set of integers in their natural order, and $T = \langle t \rangle$ the group of automorphisms of $\mathbb{Z}$ where $t : n \longmapsto n + 1$. We can construct the wreathed triangular power

$$\nabla (A, H)^{\mathbb{Z}} \text{ wr } T = (A^{(\mathbb{Z})}, \Phi \rightthreetimes H^{(\mathbb{Z})} \rightthreetimes T).$$

Take the group algebra KT and consider the more extensive representation

$$(A^{(\mathbb{Z})} \oplus KT, \Phi \rightthreetimes H^{(\mathbb{Z})} \rightthreetimes T); \tag{14}$$

where $\Phi \rightthreetimes H^{(\mathbb{Z})}$ acts on KT trivially, but T regularly. The representation (14) is uniquely determined by $\varrho = (A, H)$; so let us denote it by $f(\varrho)$. Now we set

$$\varrho_0 = \varrho, \quad \varrho_1 = f(\varrho), \quad \varrho_{\alpha+1} = f(\varrho_\alpha).$$

Assuming ϱ_α to be naturally embedded into $\varrho_{\alpha+1}$, set

$$\varrho_\beta = \bigcup_{\alpha<\beta} \varrho_\alpha$$

for any limit ordinal β. Then the representation ϱ_λ satisfies all the necessary conditions. For the details see [73].

Appendix

TRIANGULAR PRODUCTS IN RELATED CATEGORIES

The construction of triangular product is universal enough to be easily transferred from the category of group representations to certain other related categories. In this brief appendix we restrict ourselves to the following. We define triangular products of representations of associative algebras, Lie algebras and semigroups, triangular products of linear automata; then we note (mainly without proofs) several typical results and point out several applications.

First it should be noted that the triangular product of representations of algebras or semigroups (and even of sets!) can be defined verbatim as in the group case on the basis of the category-theoretic definition from Section 3. Indeed, if ρ_1 and ρ_2 are two representations (of algebras or semigroups) then, by analogy with Section 3, one can introduce the category $\mathcal{E}(\rho_1, \rho_2)$ of faithful s-extensions of ρ_1 by ρ_2. The triangular product $\rho_1 \nabla \rho_2$ is defined then as the universal object of this category.

Clearly, however, that such a definition is not convenient for a concrete work, so that in what follows we prefer to give constructive definitions of ∇-constructions.

<u>Representations of associative and Lie algebras.</u> Let K be an arbitrary but fixed commutative ring with identity. We shall first deal with representations $\rho = (A, R)$ where A is a K-module, but R an associative K-algebra acting on A. Consider two representations $\rho_1 = (A, R_1)$ and $\rho_2 = (B, R_2)$ of this kind and define their triangular product. Let $V = A \oplus B$ and let $(A, \bar{R}_1)$ and $(B, \bar{R}_2)$ be the faithful images of ρ_1 and ρ_2 respectively. The algebra $\bar{R}_1 \oplus \bar{R}_2$ can be naturally regarded as a subalgebra

of End V. Denote by Φ the annihilator of the series $0 \subseteq A \subseteq V$ in End V. It is easy to verify that Φ is a subalgebra of End V (evidently $\Phi^2 = 0$ and $\Phi \cong \text{Hom}(B, A)$, where Hom (B, A) is considered as an algebra with zero multiplication) and that Φ and $\bar{R}_1 \oplus \bar{R}_2$ form a semidirect sum $\Phi \lambda (\bar{R}_1 \oplus \bar{R}_2)$ in End V. There arises a faithful representation $(V, \Phi \lambda (\bar{R}_1 \oplus \bar{R}_2))$ which, by means of the natural epimorphism $R_1 \oplus R_2 \twoheadrightarrow \bar{R}_1 \oplus \bar{R}_2$, can be lifted up to the representation

$$(V, \Phi \lambda (R_1 \oplus R_2)) = (A \oplus B, \text{Hom}(B, A) \lambda (R_1 \oplus R_2)).$$

The latter is called the <u>triangular product of ρ_1 and ρ_2</u> and is denoted by

$$\rho_1 \nabla \rho_2 = (A, R_1) \nabla (B, R_2).$$

From the definition, it is not hard to deduce that the action in $\rho_1 \nabla \rho_2$ satisfies the following condition: for any $a \in A$, $b \in B$, $\varphi \in \Phi$, $r_1 \in R_1$, $r_2 \in R_2$

$$(a + b) \circ (\varphi + r_1 + r_2) = b\varphi + a \circ r_1 + b \circ r_2.$$

All the basic properties of triangular products of group representations from Section 2 and 3 remain true for representations of associative algebras. Moreover, the greater part of the results of these notes can be transferred, almost literally, to representations of algebras (see, for example, [26] and [75]). We believe it is not worthwile repeating here all these results; let us mention only several applications to varieties.

Following [35], recall several definitions. Let $\mathfrak{F} = K\langle x_1, x_2, \dots \rangle$ be the free associative K-algebra of countable rank. An element $f \in \mathfrak{F}$ is called an <u>identity</u> of a representation (A, R) if for every homomorphism $\mu : \mathfrak{F} \to R$ the element f^μ annihilates A. A class of representations of algebras is called a <u>variety</u> if it is determined by some set of identities. All the identities of a given variety form a T-ideal of $\mathfrak{F}$, i.e. an ideal which is invariant under all endomorphisms of $\mathfrak{F}$. If $\mathfrak{X}$ and $\mathfrak{Y}$ are varieties, then their <u>product</u> $\mathfrak{X}\mathfrak{Y}$ consists of all representations (A, R) such that A possesses an R-submodule B with $(B, R) \in \mathfrak{X}$ and $(A/B, R) \in \mathfrak{Y}$. Under this multiplication, the set of all varieties over a given K forms a <u>semigroup</u> which is anti-isomorphic to the semigroup of all T-ideals of $\mathfrak{F}$. By analogy with the theory of varieties of group representations, one can define other standard notions.

It turns out that an analogue of Theorem 6.2 is valid for representations of algebras: if ρ_1 and ρ_2 are projective representations over an integral domain K and var ρ_2 is a projective variety, then

$$\text{var}\ (\rho_1 \nabla \rho_2) = \text{var}\ \rho_1 \cdot \text{var}\ \rho_2; \tag{1}$$

the proof follows verbatim that of 6.2(if K is a field, this was established in [25]). As in the case of group representations, this theorem has a number of useful applications. For instance, it allows to prove that over any field the semigroup of varieties of representations of algebras is free [25]. In other words, we obtain another proof of the following well known.

THEOREM (Bergman and Lewin [3]). The semigroup of T-ideals of a free associative algebra over a field is free.

Using (1), it is easy to describe the identities of the full triangular matrix algebra over an arbitrary integral domain K. Indeed, let now $T_n(K)$ denotes the algebra of triangular $n \times n$ matrices over K, and let $(K^n, T_n(K))$ be its natural representation in K^n. It is evident that

$$(K^n, T_n(K)) = (K, K)\nabla \dots \nabla (K, K) \tag{2}$$

where (K, K) is the regular representation. Repeating the arguments of Section 7, we deduce from (1) and (2) that

$$\text{var}\ (K^n, T_n(K)) = (\text{var}\ (K, K))^n. \tag{3}$$

Evidently the identities of any faithful representation (A, R) coincide with the identities of the algebra R. Therefore it follows from (3) that the ideal of identities of the algebra $T_n(K)$ is equal to I^n where I is the ideal of identities of the K-algebra K. In other words, var $(T_n(K)) = \mathfrak{N}_n \cdot$ var K where $\mathfrak{N}_n$ is the variety of n-nilpotent algebras.

It is well known that if K is an infinite domain, then I is generated (as a

T-ideal) by the identity $[x_1, x_2] = x_1x_2 - x_2x_1$, but if $|K| = m < \infty$, I is generated by $x^m - x$. This completes the description of the identities of the algebra $T_n(K)$. For instance, if K is an infinite integral domain, then it is now clear that the ideal of identities of $T_n(K)$ is generated by an element

$$x_1, x_2 \quad x_3, x_4 \cdots x_{2n-1}, x_{2n}.$$

NOTE. The identities of the algebra of triangular matrices were described by several authors. First, Yu.N.Mal'cev [42] described the identities of $T_n(K)$ over a field K of characteristic 0 and raised a question of describing them over an arbitrary field [8]. This question was independently solved by Kal'julaid [26] and P.Siderov; the proof of [26] is based on triangular products. Finally, Polin [59] has recently described the identities of $T_n(K)$ over a more extensive class of rings (containing all integral domains).

Now we shall deal with Lie representations, that is, with pairs (A, L) where A is a module over a commutative ring K with 1, but L a Lie algebra over K for which there is given a representation $\circ$ in A. The triangular product of Lie representations is defined in complete analogy with the corresponding constructions for representations of groups and associative algebras. Namely, let (A, L_1) and (B, L_2) be two Lie representations. Consider their direct sum $(A \oplus B, L_1 \oplus L_2)$ and let $\Phi = \mathrm{Hom}_K(B, A)$, regarded as an abelian Lie algebra. Then $L_1 \oplus L_2$ acts naturally on Φ by the rule

$$\forall b \in B: \quad b(\varphi\circ(l_1 + l_2)) = (b\varphi)\circ l_1 - (b\circ l_2)\varphi .$$

This yields a semidirect sum (= split extension) of Lie algebras $L = \Phi \leftthreetimes (L_1 \oplus L_2)$. Define now an action of L on $V = A \oplus B$:

$$(a + b)\circ(\varphi + l_1 + l_2) \overset{\text{def.}}{=\!=} b\varphi + a\circ l_1 + b\circ l_2 .$$

It is easily verified that this rule gives a Lie representation (V, L) called the triangular product of (A, L_1) and (B, L_2) and denoted by $(A, L_1) \triangledown (B, L_2)$.

The following connection between triangular products of representations of associative algebras and those of Lie representations is straightforward. For a representation (A, R) of an associative algebra R denote by $(A, R)^L$ the corresponding Lie representation (where R is considered as a Lie algebra under the commutation $[x, y] = xy - yx$). Then for any (A, R) and (B, S) we have

$$((A, R) \triangledown (B, S))^L = (A, R)^L \triangledown (B, S)^L.$$

Most of the results of the present work hold for representations of Lie algebras as well. For example, it is proved that the semigroup of varieties of Lie representations is free [38], or the corresponding theorems concerning the semigroups of completely decomposable prevarieties and radical classes of Lie representations (see the end of [75]).

Furthermore, using the methods of Section 7, Lipyanskiĭ [39] has described the identities of the triangular matrix Lie algebra over a field. Namely, denote by $T_n^L(K)$ the Lie algebra of triangular $n \times n$ matrices over a field K.

THEOREM. The ideal of identities of $T_n^L(K)$ is generated by the following identity:

$$[[x_1, x_2], [x_3, x_4], \dots, [x_{2n-1}, x_{2n}]] \quad \text{if } |K| = \infty,$$

$$[x_1^m - x_1, x_2^m - x_2, \dots, x_n^m - x_n] \quad \text{if } |K| = m < \infty$$

(the commutators are left-normed). The second element can be naturally rewritten in the "Lie" form; see 39.

We are sure that this theorem remains true over an arbitrary integral domain. We also believe that the above technique makes it possible to describe the identities of the Jordan algebra of triangular $n \times n$ matrices over any domain with $\frac{1}{2}$. Of course, one should beforehand define the triangular product of Jordan representations and establish an analogue of Theorem 6.2.

Representations of semigroups and linear automata. In contrast to representations of groups and algebras, automata are algebraic structures with three ground sets. Therefore in the theory of linear automata (and, in particular, in questions connected with triangular products) there arises a number of substantially new difficulties. The purpose of this section is very modest: to define the triangular product of linear automata and to indicate a few applications to decomposition problems. The possibility of applying the ∇-product to these problems was pointed out by Plotkin, and from now on we follow his paper [57].

By an automaton we always mean a semigroup automaton, that is, a five-tuple $\mathfrak{A} = (A, S, B, \circ, *)$ where A is a set of internal states, S a semigroup of inputs, B a set of outputs, $\circ : A \times S \to A$ a transition function and $* : A \times S \to B$ an output function. Every input s transforms a state a in a new state $a' = a \circ s$ and simultaneously s transforms a in an output $b = a * s$. Besides, the maps $\circ$ and $*$ satisfy the conditions

$$\forall\ a \in A,\ s_1 s_2 \in S : a \circ (s_1 s_2) = (a \circ s_1) \circ s_2,\ \ a * (s_1 s_2) = (a \circ s_1) * s_2.$$

As usual in such cases we shall denote an automaton simply by $\mathfrak{A} = (A, S, B)$ omitting the symbols $\circ$ and $*$.

If $\mathfrak{A} = (A, S, B)$ and $\mathfrak{A}' = (A', S', B')$ are two automata, then a morphism $\mu : \mathfrak{A} \to \mathfrak{A}'$ is a triple $\mu = (\mu_1, \mu_2, \mu_3)$ where $\mu_1 : A \to A'$ and $\mu_3 : B \to B'$ are maps but $\mu_2 : S \to S'$ a homomorphism of semigroups for which the following conditions are satisfied:

$$\forall\ a \in A,\ s \in S : (a \circ s)^{\mu_1} = a^{\mu_1} \circ s^{\mu_2},\ \ (a * s)^{\mu_3} = a^{\mu_1} * s^{\mu_2}.$$

Clearly we obtain the category of automata.

A semiautomaton, or an action, is a pair (A, S) consisting of a set A and a semigroup S for which there is given an action on A, that is, a map $\circ : A \times S \to A$ satisfying the condition

$$\forall\ a \in A,\ s_1 s_2 \in S : a \circ (s_1 s_2) = (a \circ s_1) \circ s_2.$$

Evidently any automaton $\mathfrak{A} = (A, S, B)$ includes a semiautomaton (A, S) called the base semiautomaton of $\mathfrak{A}$. The category of semiautomata is defined in a natural way, and it is clear that the transition from an automaton to its base semiautomaton is a functor.

All the above relates to "pure automata, i.e. to automata whose sets A and B are not provided with any supplementary structure. Consider now a more familiar situation. Let K be an arbitrary commutative ring with identity. An automaton (A, S, B) is called a linear automaton over K if A and B are K-modules and for each $s \in S$ the maps $a \longmapsto a \circ s$ and $a \longmapsto a * s$ are K-linear. It is evident how to define morphisms of linear automata, semiautomata, etc. If (A, S, B) is a linear automata, then its base semiautomaton (A, S) is actually a usual representation of a semigroup S in a module A. Now it will be convenient to say that (A, S) is a linear semiautomaton over K. The category of linear semiautomata over K (= the category of semigroup representations over K) is defined in a standard way, and it is clear that it can be considered as a subcategory of the category of linear K-automata.

One of the major results in the modern algebraic theory of (pure) automata is the Krohn-Rhodes Decomposition Theorem asserting that every semiautomaton can, in a certain sense, be constructed from certain "simple" semiautomata. Nothing similar has been known in the theory of linear automata for a long time. However, it has been recently discovered that the construction of triangular product of linear automata can be successfully used for this purpose. To define this construction, consider first two linear semiautomata (A_1, S_1) and (A_2, S_2) over K. Their triangular product is defined in the usual manner, so that we omit the details. Note only that

$$(A_1, S_1) \nabla (A_2, S_2) = (A_1 \oplus A_2, \Phi \leftthreetimes (S_1 \times S_2))$$

where $\Phi = \mathrm{Hom}\,(A_2, A_1)$ is regarded as an additive group.

Now let there are given two linear automata $\mathfrak{A}_1 = (A_1, S_1, B_1)$ and $\mathfrak{A}_2 = (A_2, S_2, B_2)$. Consider the additive groups $\Phi = \mathrm{Hom}\,(A_2, A_1)$ and $\Psi = \mathrm{Hom}\,(A_2, B_1)$. Define left actions of the semigroup S_2 on Φ and Ψ by the rules: if $a \in A_2$, $s_2 \in S_2$, $\varphi \in \Phi$ and $\psi \in \Psi$, then

$$a(s_2 \circ \varphi) = (a \circ s_2)\varphi \;, \quad a(s_2 \circ \psi) = (a \circ s_2)\psi \;.$$

Define a right action of S_1 on Φ : if $a \in A_2$ and $s_1 \in S_1$, then

$$a(\varphi \circ s_1) = (a\varphi)\circ s_1.$$

Furthermore, for any $\varphi \in \Phi$ and $s_1 \in S_1$ define an element $\varphi * s_1 \in \Psi$ by the rule: if $a \in A_2$, then

$$a(\varphi * s_1) = (a\varphi) * s_1.$$

It is easily verified that all these actions agree with linear operations on Φ and Ψ and that $\varphi * (s_1 s_1') = (\varphi \circ s_1) * s_1'$ for all $\varphi \in \Phi$, $s_1, s_1' \in S_1$. Take now the Cartesian product of sets $S = \Phi \times \Psi \times S_1 \times S_2$ and define a multiplication on S by

$$(\varphi, \psi, s_1, s_2)(\varphi', \psi', s_1', s_2') = (\varphi \circ s_1' + s_2 \circ \varphi', \varphi * s_1' + s_2 \circ \psi', s_1 s_1', s_2 s_2').$$

The S becomes a semigroup. Finally, denote $A = A_1 \oplus A_2$, $B = B_1 \oplus B_2$ and for any $a = a_1 + a_2 \in A$, $s = (\varphi, \psi, s_1, s_2) \in S$ set

$$a\circ s = a_1 \circ s_1 + a_2 \varphi + a_2 \circ s_2, \quad a * s = a_1 * s_1 + a_2 \psi + a_2 * s_2.$$

One can verify that we obtain a linear automata $\mathfrak{A} = (A, S, B)$ called the <u>triangular product of $\mathfrak{A}_1$ and $\mathfrak{A}_2$</u> and denoted by $\mathfrak{A} = \mathfrak{A}_1 \triangledown \mathfrak{A}_2$. It is clear that the base semiautomaton (A, S) of $\mathfrak{A}$ is just the triangular product of the base semiautomata of $\mathfrak{A}_1$ and $\mathfrak{A}_2$.

We shall need one simple notion. Let $\mathfrak{A}_1$ and $\mathfrak{A}_2$ be two objects of some category. $\mathfrak{A}_1$ is a <u>divisor</u> (or a <u>factor</u>) of $\mathfrak{A}_2$ if it is an epimorphic image of some subobject of $\mathfrak{A}_2$.

We can now state a few results on decomposition of automata. Recall first the formulation of the Krohn-Rhodes Theorem. It asserts that every (pure) finite semiautomaton is a divisor of the wreath product of semiautomata $\mathfrak{A}_1, .., \mathfrak{A}_n$ such that every $\mathfrak{A}_i$ is either a trigger (i.e. a simplest semiautomaton with two states) or a semiautomaton whose active semigroup is a simple group ([29] ; see also [2] and [10]). This fact is actually a far-going generalization of the Kaloujnine-Krasner Theorem according to which every finite group is embedded in the wreath product of its simple divisors.

Now let us turn to linear automata. In this case certain analogous results have been recently obtained in a paper of Finkelstein [11] . Namely, to any linear semiautomaton (A, S) over K one can naturally assign a representation (A, KS) of a semigroup algebra KS. It turns out that, by means of this correspondence, the decomposition theory of linear semiautomata is reduced to the decomposition of representations of associative algebras into triangular products.

Two cases are considered in [11] : K is a field and $K = \mathbb{Z}$. In the first case it is proved that every finite-dimensional representation (A, R) is a divisor of the triangular product $(A_1, R_1)\nabla\ldots\nabla(A_n, R_n)$ where all (A_i, R_i) are divisors of (A, R); moreover, each (A_i, R_i) is <u>simple</u>, that is, it cannot be represented in such a form. Furthermore, it is proved that a representation is simple if and only if it is faithful and irreducible.

In the case $K = \mathbb{Z}$ an analogous theorem is established: if (A, R) is a representation with finitely generated abelian group A, then it is a divisor of the triangular product of its simple divisors. Simple representations are also completely described: (A, R) is simple if and only if either (i) R is the full matrix ring over a Galois ring and A a faithful indecomposable R-module, or (ii) A is a free abelian group without R-submodules of lesser $\mathbb{Z}$-rank.

Unfortunately, all these results (as well as Krohn-Rhodes' theory) relate only to semiautomata, i.e. to automata without outputs. As for "full" automata, the following is known. Every finite-dimensional automaton (A, S, B) over a field is monomophically embedded into the triangular product of its divisors (A_i, S_i, B_i) where (A_i, S_i) are irreducable but $\dim B_i = 1$. The proof is not complicated: it is based on an analogue of the Embedding Theorem from Section 3. It is considerably more complicated to prove that (A_i, S_i, B_i) can be reduced to automata of the kind (A'_j, S'_j, B'_j), where each (A'_j, S'_j) is an irreducible representation of a group S'_j, and to pure triggers. These results are still in a preliminary form and we omit their precise formulations. For details we refer the reader to forthcoming papers of Plotkin et al.

REFERENCES

1. Ado, I.D., On nilpotent algebras and p-groups, DAN SSSR 40 (1943), 299-301.

2. Algebraic theory of machines, languages and semigroups, Edited by M.A.Arbib, Academic Press, 1968.

3. Bergman, G.M. and Lewin, J., The semigroup of ideals of a fir is (usually) free, J. London Math. Soc. 11 (1975), 21-31.

4. Birkhoff, G., Lattice theory, 3rd ed., A.M.S., Providence, 1967.

5. Buckley, J., Polynomial functions and wreath products, Illinois J. Math. 14 (1970), 274-282.

6. Bunt, A.Ya., Automorphisms of triangular products, Collection of Papers in Algebra, Riga, 1978, 3 - 8.

7. Cohn, P.M., Free ideal rings, J.Algebra 1 (1964), 47 - 69.

8. Dnestr. Notebook, Institute of Mathematics, Novosibirsk, 1976.

9. Dunwoody, M.J., On product varieties, Math.Z. 104 (1968), 91 - 97.

10. Eilenberg, S., Automata, languages and machines, vol. B, Academic Press, 1975.

11. Finkelstein, M.Ya., Decomposition of linear automata, Abelian Groups and Modules, Tomsk State Univ., Tomsk, 1980, 109-125.

12. Fox, R.H., Free differential calculus, I, Ann. Math..57 (1953), 547-560.

13. Grinberg, A.S., Varieties of stable linear representations, Latv. Mat. Yežegodnik 9 (1971), 39-46.

14. Gringlaz, L.Ya. and Plotkin, B.I., A note on powers of varieties of pairs and faithful group representations, Latv. Mat. Yežegodnik 16 (1974), 23-32.

15. Grienberg, K.W., The residual nilpotence of certain presentations of finite groups, Arch. Math. 13 (1962), 408-417.

16. Gruenberg, K.W. and Roseblade, J.E., The augmentation terminals of certain locally finite groups, Can. J. Math. 24 (1972), 221-338.

17. Hall, P., Nilpotent groups, Canad. Math. Congress Summer Sem., Univ. of Alberta, Edmonton, 1957.

18. Hall, P., Wreath powers and characteristically simple groups, Proc. Cambr. Phil.Soc. 58 (1962), 170-184.

19. Hall, P., On non-strictly simple groups, Proc.Cambr.Phil.Soc. 59 (1963), 531-553.

20. Hartley, B., The residual nilpotence of wreath products, Proc. London Math. Soc. 20 (1970), 365-392.

21. Jennings, S.A., The structure of the group ring of a p-group over a modular field, Trans. Amer. Math. Soc. 50 (1941), 175-185.

22. Jennings, S.A., The group ring of a class of infinite nilpotent groups, Can. J. Math. 7 (1955), 169-187.

23. Jónsson, B., On the representations of lattices, Math. Scand. 1 (1953), 193-206.

24. Kal'julaid, U.E., On the powers of the augmentation ideal, Eesti NSV Teaduste Akademia Toimetised 22 (1973), 3-21.

25. Kal'julaid, U.E., Triangular products of representations of semigroups and associative algebras, Uspehi Mat. Nauk 32, No.4 (1977), 253-254.

26. Kal'julaid, U.E., Triangular products and stability of representations, Thesis, Tartu State University, Tartu, 1978.

27. Kaloujnine, L., Uber gewisse Beziehungen zwischen einer Gruppe und ihren Automorphismen, Berliner Math. Tagung (1953), 164-172.

28. Kaloujnine, L. and Krasner, M., Produit complete des groupes de permutations et le probleme d'extension des groupes III, Acta Sci. Math. (Szeged) 14 (1951), 69-82.

29. Krohn, K. and Rhodes, J., Algebraic theory of machines. I. Prime decomposition theorem for finite semigroups and machines, Trans. Amer. Math. Soc. 116 (1965), 450-464.

30. Krop, L.E., Solvable varieties of pairs, Latv. Mat. Yežegodnik 18 (1976), 64-80.

31. Krop, L.E. and Plotkin, B.I., Magnus varieties of group representations, Mat. Sb. 95 (1974), 499-524.

32. Krop, L.E. and Simonjan, L.A., On pure and Magnus varieties of group representations, Collection of Papers in Algebra, Riga, 1978, 108-123.

33. Krull, W., Uber verallgemeinerte endlische Abelsche Gruppen, Math. Z. 23 (1925), 161-196.

34. Kublanova, E.M. and Plotkin, B.I., On the algebra of varieties of group representations over an arbitrary commutative ring, Latv.Mat.Yežegodnik 24 (1980), 183-196.

35. Kuros, A.G., The theory of groups, 3rd ed., Nauka, 1967.

36. Kuškulei, A.H., Of finitely stable representations of nilpotent groups, Latv. Mat. Yežegodnik 16 (1975), 39-45.

37. Lazard, M., Sur les groupes nilpotents et les anneaux de Lie, Ann. Sci. Ecole Norm. sup. 71 (1954), 101-190.

38. Lipyanskiǐ, R.S., The semigroup of varieties of Lie pairs, Set Theory and Topology 1, Udmurd. State Univ., Iževsk, 1977, 44-54.

39. Lipyanskiǐ, R.S., On identities of the variety generated by triangular matrices, Topological Spaces and Maps 4, Latv. State. Univ., Riga, 1979, 74-76.

40. Magnus, W., Beziehungen zwischen Gruppen und Idealen in einem speziellen Ring, Math. Ann. 111 (1935), 259-280.

41. Mal'cev, A.I., Generalized nilpotent algebras and their adjoint groups, Mat. Sb. 25 (1949), 347-366.

42. Mal'cev, Yu.N., A basis for identities of the algebra of upper triangular matrices, Algebra i Logika 10 (1971), 393-400.

43. McLain, D.H., A characteristically simple group, Proc. Cambr. Phil. Soc. 50 (1954), 641-642.

44. Neumann, B.H., Neumann, H. and Neumann, P.M., Wreath products and varieties of groups, Math. Z. 80 (1962), 44-62.

45. Neumann, H., Varieties of groups, Springer-Verlag, 1967.

46. Neumann, P.M., On the structure of standard wreath products of groups, Math. Z. 84 (1964), 343-373.

47. Passi, I.B.S., Group rings and their augmentation ideals, Lect. Notes Math. 715, Springer-Verlag, 1979.

48. Plotkin, B.I., On the semigroup of radical classes of groups, Sibirsk. Mat. Ž. 10 (1969), 1091-1108.

49. Plotkin, B.I., Functorials, radicals and coradicals in groups, Matem. zap. Ural. Gos. Univ. 7, No.3 (1970), 150-182.

50. Plotkin, B.I., Triangular products of pairs, Certain Questions of Group Theory, Latv. Valsts Univ. Zinatn. Raksti 151 (1971), 140-170.

51. Plotkin, B.I., Radicals and varieties in group representations, Latv. Mat. Yežegodnik 10 (1972), 75-132.

52. Plotkin, B.I., Group varieties and varieties of pairs associated with group representations, Sibirsk. Mat. Ž. 13 (1972), 1030-1053.

53. Plotkin, B.I., Notes on stable representations of nilpotent groups, Trans. Moskov. Math. Soc. 29 (1973), 185-200.

54. Plotkin, B.I., Multiplicative systems of varieties of pairs - group representations, Latv. Mat. Yežegodnik 18 (1976), 143-170.

55. Plotkin, B.I., Varieties of group representations, Uspehi Mat. Nauk 32, No.5, (1977), 3-68.

56. Plotkin, B.I., Locally finite and locally finite-dimensional varieties of pairs - group representations, Collection of Papers in Algebra, Riga, 1978, 188-245.

57. Plotkin, B.I., Algebraic models of automata. Certain constructions and problems (preprint).

58. Plotkin, B.I. and Grinberg, A.S., Semigroups of varieties associated with group representations, Sibirsk. Mat. Ž. 13 (1972), 841-858.

59. Polin, S.V., Identities of the algebra of triangular matrices, Sibirsk. Mat. Ž. 21 (1980), 206-215.

60. Rips, E., On the fourth integer dimension subgroup, Israel J. Math. 12 (1972), 342-346.

61. Romanovskiĭ, N.S., Bases for identities of certain matrix groups, Algebra i Logika 10 (1971), 401-406.

62. Schmidt, O.Yu., Uber unendlische Gruppen mit endlischer Kette, Math. Z. 29 (1928), 34-41.

63. Silcock,H.L., Generalized wreath products and the lattice of normal subgroups of a group, Algebra Univers. 7 (1977), 361-372.

64. Smith, P.E., On the intersection theorem, Proc. London Math. Soc. 21 (1970), 385-396.

65. Šmel'kin, A.L., The semigroup of varieties of groups, DAN SSSR 149 (1963), 543-545.

66. Tsalenko, M.S., The "semigroup" of reflective subcategories, Mat. Sb. 81 (1970), 62-78.

67. Vovsi, S.M., Absolute freeness of nonstrict radicals, Certain Questions of Group Theory, Latv. Valsts Univ. Zinatn. Raksti 151 (1971), 14-18.

68. Vovsi, S.M., On infinite products of classes of groups, Sibirsk. Mat. Ž. 13 (1972), 272-285.

69. Vovsi, S.M., The semigroup of prevarieties of linear group representations, Mat. Sb. 93 (1974), 405-421.

70. Vovsi, S.M., Triangular products, prevarieties and radical classes of linear group representations, DAN SSSR 224 (1975), 27-30.

71. Vovsi, S.M., Triangular products and semigroups of classes of pairs, Latv. Mat. Yežegodnik 18 (1976), 27-39.

72. Vovsi, S.M., The isomorphism of triangular decompositions of group representations, Latv. Mat. Yežegodnik 21 (1977), 116-123.

73. Vovsi, S.M., Infinite triangular products of group representations, Math. Slovaca 27 (1977), 337-358.

74. Vovsi, S.M., Prevarieties of generalized stable groups, Sibirsk. Mat. Ž. 19 (1978), 267-279.

75. Vovsi, S.M., The semigroups of radicals classes of group representations and representations of associative algebras, Collection of Papers in Algebra, Riga, 1978, 332-340.

76. Vovsi, S.M., Lattices of invariant subspaces of group representations, Notices Amer. Math. Soc. 26 (1979), A - 368; Algebra Univers 12(1981), 221-223.

77. Vovsi, S.M., On pure varieties of group representations, Notices Amer. Math. Soc. 26 (1979), A - 507.

78. Vovsi, S.M., Varieties and triangular products of projective representations of groups, Sibirsk. Mat. Ž. 21 (1980), 56-62.

79. Vovsi, S.M., A note on the ω-th integer dimension subgroup, Abstr. Amer. Math. Soc. 1 (1980), 619.

80. Wielandt, H., Eine Verallgemeinerung der invarinten Untergruppen, Math. Z. 45 (1939), 209-244.

Progress in Mathematics
Edited by J. Coates and S. Helgason

Progress in Physics
Edited by A. Jaffe and D. Ruelle

- A collection of research-oriented monographs, reports, notes arising from lectures or seminars
- Quickly published concurrent with research
- Easily accessible through international distribution facilities
- Reasonably priced
- Reporting research developments combining original results with an expository treatment of the particular subject area
- A contribution to the international scientific community: for colleagues and for graduate students who are seeking current information and directions in their graduate and post-graduate work.

Manuscripts

Manuscripts should be no less than 100 and preferably no more than 500 pages in length.

They are reproduced by a photographic process and therefore must be typed with extreme care. Symbols not on the typewriter should be inserted by hand in indelible black ink. Corrections to the typescript should be made by pasting in the new text or painting out errors with white correction fluid.

The typescript is reduced slightly (75%) in size during reproduction; best results will not be obtained unless the text on any one page is kept within the overall limit of 6x9½ in (16x24 cm). On request, the publisher will supply special paper with the typing area outlined.

Manuscripts should be sent to the editors or directly to:
Birkhäuser Boston, Inc., P.O. Box 2007, Cambridge, Massachusetts 02139

PROGRESS IN MATHEMATICS

Already published

PM 1 Quadratic Forms in Infinite-Dimensional Vector Spaces
Herbert Gross
ISBN 3-7643-1111-8, 431 pages, $22.00, paperback

PM 2 Singularités des systèmes différentiels de Gauss-Manin
Frédéric Pham
ISBN 3-7643-3002-3, 339 pages, $18.00, paperback

PM 3 Vector Bundles on Complex Projective Spaces
C. Okonek, M. Schneider, H. Spindler
ISBN 3-7643-3000-7, 389 pages, $20.00, paperback

PM4 Complex Approximation, Proceedings, Quebec, Canada, July 3-8, 1978
Edited by Bernard Aupetit
ISBN 3-7643-3004-X, 128 pages, $10.00, paperback

PM 5 The Radon Transform
Sigurdur Helgason
ISBN 3-7643-3006-6, 202 pages, $14.00, paperback

PM 6 The Weil Representation, Maslov Index and Theta Series
Gérard Lion, Michèle Vergne
ISBN 3-7643-3007-4, 345 pages, $18.00, paperback

PM 7 Vector Bundles and Differential Equations
Proceedings, Nice, France, June 12-17, 1979
Edited by André Hirschowitz
ISBN 3-7643-3022-8, 255 pages, $16.00, paperback

PM 8 Dynamical Systems, C.I.M.E. Lectures, Bressanone, Italy, June 1978
John Guckenheimer, Jürgen Moser, Sheldon E. Newhouse
ISBN 3-7643-3024-4, 300 pages, $16.00, paperback

PM 9 Linear Algebraic Groups
T. A. Springer
ISBN 3-7643-3029-5, 304 pages, $18.00, hardcover

PM10 Ergodic Theory and Dynamical Systems I
A. Katok
ISBN 3-7643-3036-8, 352 pages, $18.00, hardcover

PM11 18th Scandinavian Congress of Mathematicians, Aarhus, Denmark, 1980
Edited by Erik Balslev
ISBN 3-7643-3034-6, 528 pages, $26.00, hardcover

PM12 Séminaire de Théorie des Nombres, Paris 1979-80
Edited by Marie-José Bertin
ISBN 3-7643-3035-X, 408 pages, $22.00, hardcover

PM13 Topics in Harmonic Analysis on Homogeneous Spaces
Sigurdur Helgason
ISBN 3-7643-3051-1, 142 pages, $12.00, hardcover

PM14 Manifolds and Lie Groups, Papers in Honor of Yozô Matsushima
Edited by J. Hano, A. Marimoto, S. Murakami, K. Okamoto, and H. Ozeki
ISBN 3-7643-3053-8, 480 pages, $35.00, hardcover

PM15 Representations of Real Reductive Lie Groups
David A. Vogan, Jr.
ISBN 3-7643-3037-6, 771 pages, $35.00, hardcover

PM16 Rational Homotopy Theory and Differential Forms
Phillip A. Griffiths, John W. Morgan
ISBN 3-7643-3041-4, 264 pages, $16.00, hardcover

PROGRESS IN PHYSICS

Already published

PPh1 Iterated Maps on the Interval as Dynamical Systems
Pierre Collet and Jean-Pierre Eckmann
ISBN 3-7643-3026-0, 256 pages, $16.00 hardcover

PPh2 Vortices and Monopoles, Structure of Static Gauge Theories
Arthur Jaffe and Clifford Taubes
ISBN 3-7643-3025-2, 275 pages, $16.00 hardcover

PPh3 Mathematics and Physics
Yu. I. Manin
ISBN 3-7643-3027-9, 111 pages, $10.00 hardcover